Artificial Unintelligence

Artificial Unintelligence

How Computers Misunderstand the World

Meredith Broussard

The MIT Press
Cambridge, Massachusetts
London, England

This book was set in ITC Stone Serif Std by Toppan Best-set Premedia Limited. Printed and bound in the United States of America.

Library of Congress Cataloging-in-Publication Data

Names: Broussard, Meredith, author.
Title: Artificial unintelligence : how computers misunderstand the world / Meredith Broussard.
Description: Cambridge, MA : MIT Press, [2018] | Includes bibliographical references and index.
Identifiers: LCCN 2017041363 | ISBN 9780262038003 (hardcover : alk. paper)
Subjects: LCSH: Electronic data processing--Social aspects. | Computer programs--Correctness. | Errors.
Classification: LCC QA76.9.C66 B787 2018 | DDC 303.48/34--dc23 LC record available at https://lccn.loc.gov/2017041363

10 9 8 7 6 5 4 3 2 1

For my family

Contents

I   How Computers Work   1

1   Hello, Reader   3
2   Hello, World   13
3   Hello, AI   31
4   Hello, Data Journalism   41

II   When Computers Don't Work   49

5   Why Poor Schools Can't Win at Standardized Tests   51
6   People Problems   67
7   Machine Learning: The DL on ML   87
8   This Car Won't Drive Itself   121
9   Popular Doesn't Mean Good   149

III   Working Together   161

10   On the Startup Bus   163
11   Third-Wave AI   175
12   Aging Computers   193

Acknowledgments   201
Notes   203
Bibliography   211
Index   227

I   How Computers Work

1 Hello, Reader

I love technology. I've loved it since I was a little girl and my parents bought me an Erector Set that I used to build a giant (to me) robot out of small pierced pieces of metal. The robot was supposed to be powered by a miniature, battery-driven motor. I was an imaginative kid; I convinced myself that once this robot was built, it would move around the house as easily as I did, and I would have a new robot best friend. I would teach the robot to dance. It would follow me around the house, and (unlike my dog) it would play fetch.

I spent hours sitting on the red wool rug in the upstairs hallway of my parents' house, daydreaming and assembling the robot. I tightened dozens of nuts and bolts using the set's tiny, child-sized wrenches. The most exciting moment came when I was ready to plug in the motor. My mom and I made a special trip to the store to get the right batteries for the motor. We got home, and I raced upstairs to connect the bare wires to the gears and turn on my robot. I felt like Orville and Wilbur Wright at Kitty Hawk, launching a new machine and hoping it would change the world.

Nothing happened.

I checked the diagrams. I flicked the on/off switch a few times. I flipped the batteries. Still nothing. My robot didn't work. I went to get my mom.

"You need to come upstairs. My robot isn't working," I said sadly.

"Did you try turning it off and turning it on again?" my mom asked.

"I did that," I said.

"Did you try flipping the batteries?" she asked.

"Yes," I said. I was getting frustrated.

"I'll come look at it," she said. I grabbed her hand and pulled her upstairs. She tinkered with the robot for a little bit, looking at the directions and fiddling with the wiring and turning the switch on and off a few times. "It's not working," she said finally.

"Why not?" I asked. She could have just told me that the motor was broken, but my mother believed in complete explanations. She told me that the motor was broken, and then she also explained global supply chains and assembly lines and reminded me that I knew how factories worked because I liked to watch the videos on *Sesame Street* featuring huge industrial machines making packages of crayons.

"Things can go wrong when you make things," she explained. "Something went wrong when they made this motor, and it ended up in your kit anyway, and now we're going to get one that works." We called the Erector hotline number printed on the instructions, and the nice people at the toy company sent us a new motor in the mail. It arrived in a week or so, and I plugged it in, and my robot worked. By that point, it was anticlimactic. The robot worked, but not well. It could move slowly across the hardwood floor. It got stuck on the rug. It wasn't going to be my new best friend. After a few days, I took the robot apart to make the next project in the kit, a Ferris wheel.

I learned a few things from making this robot. I learned how to use tools to build technology, and that building things could be fun. I discovered that my imagination was powerful, but that the reality of technology couldn't measure up to what I imagined. I also learned that parts break.

A few years later, when I began writing computer programs, I discovered that these lessons from robot building translated well to the world of computer code. I could imagine vastly complex computer programs, but what the computer could *actually* do was often a letdown. I ran into many situations where programs didn't work because a part failed somewhere in the computer's innards. Nevertheless, I persisted, and I still love building and using technology. I have a vast number of social media accounts. I once hacked a crockpot to build a device for tempering twenty-five pounds of chocolate as part of a cooking project. I even built a computerized system to automatically water my garden.

Recently, however, I've become skeptical of claims that technology will save the world. For my entire adult life, I've been hearing promises about what technology can do to change the world for the better. I began studying computer science at Harvard in September 1991, months after Tim Berners-Lee launched the world's first website at CERN, the particle physics lab run by the European Organization for Nuclear Research. In my sophomore year, my roommate bought a NeXT cube, the same square black computer that Berners-Lee used as a web server at CERN. It was fun. My roommate got a high-speed connection in our dormitory suite, and we used his $5,000 computer to check our email. Another roommate, who had recently come out and was too young for Boston's gay bar scene, used the computer to hang out on online bulletin boards and meet boys. It was easy to believe that in the future, we would do everything online.

For youthful idealists of my generation, it was also easy to believe that the world we were creating online would be better and more just than the world we already had. In the 1960s, our parents thought they could make a better world by dropping out or living in communes. We saw that our parents had gone straight, and communes clearly weren't the answer—but there was this entire new, uncharted world of "cyberspace" that was ours for the making. The connection wasn't just metaphorical. The emerging Internet culture of the time was heavily influenced by the New Communalism movement of the 1960s, as Fred Turner writes in *From Counterculture to Cyberculture*, a history of digital utopianism.[1] Stewart Brand, the founder of the *Whole Earth Catalog*, laid out the connections between the counterculture and the personal computer revolution in an essay called "We Owe It All to the Hippies," in a 1995 special issue of *Time* magazine called "Welcome to Cyberspace."[2] The early Internet was deeply groovy.

By my junior year, I could make a web page or spin up a web server or write code in six different programming languages. For an undergraduate majoring in math, computer science, or engineering at the time, this was completely normal. For a woman, it wasn't. I was one of six undergraduate women majoring in computer science at a university of twenty thousand graduate and undergraduate students. I only knew two of the other women in computer science. The other three felt like a rumor. I felt isolated in all of the textbook ways that cause women to drop out of science, technology, engineering, and mathematics (STEM) careers. I could see what was broken inside the system, for me and for other women, but I didn't have the power to fix it. I switched my major.

I took a job as a computer scientist after college. My job was to make a simulator that was like a million bees with machine guns attacking all at once so that we could deploy the bees against a piece of software, to test that the software wouldn't go down when it was deployed. It was a good job, but I wasn't happy. Again, it felt like there was nobody around who looked like me or talked like me or was interested in the things that interested me. I quit to become a journalist.

Fast-forward a few years: I went back to computer science as a data journalist. *Data journalism* is the practice of finding stories in numbers and using numbers to tell stories. As a data journalist, I write code in order to commit acts of investigative journalism. I'm also a professor. It suits me. The gender balance is better, too.

Journalists are taught to be skeptical. We tell each other, "If your mother says she loves you, check it out." Over the years, I heard people repeat the same promises about the bright technological future, but I saw the digital world replicate the inequalities of the "real" world. For example, the percentage of women and minorities in the tech workforce never increased significantly. The Internet became the new public sphere, but friends and colleagues reported being harassed online more than they ever were before. My women friends who used online dating sites and apps received rape threats and obscene photos. Trolls and bots made Twitter a cacophony.

I started to question the promises of tech culture. I started to notice that the way people talk about technology is out of sync with what digital technology actually can do. Ultimately, everything we do with computers comes down to math, and there are fundamental limits to what we can (and should) do with it. I think that we have reached that limit. Americans have hit a point at which we are so enthusiastic about using technology for everything—hiring, driving, paying bills, choosing dates—that we have stopped demanding that our new technology is *good*.

Our collective enthusiasm for applying computer technology to every aspect of life has resulted in a tremendous amount of poorly designed technology. That badly designed technology is getting in the way of everyday life rather than making life easier. Simple things like finding a new friend's phone number or up-to-date email address have become time-consuming. The problem here, as in so many cases, is too much technology and not enough people. We turned over record-keeping to computational systems but fired all the humans who kept the information up-to-date. Now, since nobody goes through and makes sure all the contact information is accurate in every institutional directory, it is more difficult than ever to get in touch with people. As a journalist, a lot of my job involves reaching out to people I don't know. It's harder than it used to be, and it's more expensive, to contact anyone.

There's a saying: When all you have is a hammer, everything looks like a nail. Computers are our hammers. It's time to stop rushing blindly into the digital future and start making better, more thoughtful decisions about when and why to use technology.

Hence, this book.

This book is a guide for understanding the outer limits of what technology can do. It's about understanding the bleeding edge, where human achievement intersects with human nature. That edge is more like a cliff; beyond it lies danger.

The world is full of marvelous technology: Internet search, devices that recognize spoken commands, computers that can compete against human experts in games like Jeopardy! or Go. In celebrating these achievements, it's important not to get too carried away and assume that because we have cool technology, we can use it to solve every problem. In my university classes, one of the fundamental things I teach is that there are limits. Just as there are fundamental limits to what we know in mathematics and in science, so are there fundamental limits to what we can do with technology. There are also limits to what we *should* do with technology. When we look at the world only through the lens of computation, or we try to solve big social problems using technology alone, we tend to make a set of the same predictable mistakes that impede progress and reinforce inequality. This book is about how to understand the outer limits of what technology can do. Understanding these limits will help us make better choices and have collective conversations as a society about what we can do with tech and what we ought to do to make the world truly better for everyone.

I come to this conversation about social justice as a journalist. I specialize in a particular kind of data journalism, called *computational journalism* or *algorithmic accountability reporting*. An *algorithm* is a computational procedure for deriving a result, much like a recipe is a procedure for making a particular dish. Sometimes, algorithmic accountability reporting means writing code to investigate the algorithms that are being used increasingly to make decisions on our behalf. Other times, it means looking at badly designed technology or falsely interpreted data and raising a red flag.

One of the red flags I want to raise in this book is a flawed assumption that I call *technochauvinism*. Technochauvinism is the belief that tech is always the solution. Although digital technology has been an ordinary part of scientific and bureaucratic life since the 1950s, and everyday life since the 1980s, sophisticated marketing campaigns still have most people convinced that tech is something new and potentially revolutionary. (The tech revolution has already happened; tech is now mundane.)

Technochauvinism is often accompanied by fellow-traveler beliefs such as Ayn Randian meritocracy; technolibertarian political values; celebrating free speech to the extent of denying that online harassment is a problem; the notion that computers are more "objective" or "unbiased" because they distill questions and answers down to mathematical evaluation; and an unwavering faith that if the world just used more computers, and used them properly, social problems would disappear and we'd create a digitally enabled utopia. It's not true. There has never been, nor will there ever be, a technological innovation that moves us away from the essential problems of human nature. Why, then, do people persist in thinking there's a sunny technological future just around the corner?

I started thinking about technochauvinism one day when I was talking with a twenty-something friend who works as a data scientist. I mentioned something about Philadelphia schools that didn't have enough books.

"Why not just use laptops or iPads and get electronic textbooks?" asked my friend. "Doesn't technology make everything faster, cheaper, and better?"

He got an earful. (You'll get one too in a later chapter.) However, his assumption stuck with me. My friend thought that technology was always the answer. I thought technology was only appropriate if it was the right tool for the task.

Somehow, in the past two decades, many of us began to assume that computers get it right and humans get it wrong. We started saying things like "Computers are better because they are more objective than people." Computers have become so pervasive in every aspect of our lives that when something goes awry in the machine, we assume that it's our fault, rather than assume something went wrong within the thousands of lines of code that make up the average computer program. In reality, as any software developer can tell you, the problem is usually in the machine somewhere. It's probably in poorly designed or tested code, cheap hardware, or a profound misunderstanding of how the actual users would use the system.

If you're anything like my data scientist friend, you're probably skeptical. Maybe you're a person who loves your cell phone, or maybe you've been told your whole life that computers are the wave of the future. I hear you. I was told that too. What I ask is that you stick with me as I tell some stories about people who built technology, then use these stories to think critically about the technology we have and the people who made it. This isn't a technical manual or a textbook; it's a collection of stories with a purpose. I chose a handful of adventures in computer programming, each of which I undertook in order to understand something fundamental about technology and contemporary tech culture. All of those projects link together in a sort of chain, building an argument against technochauvinism. Along the way, I'll explain how some computer technology works and unpack the human systems that technology serves.

The first four chapters of the book cover a few basics about how computers work and how computer programs are constructed. If you already are crystal-clear on how hardware and software work together, or you already know how to write code, you'll probably breeze through chapters 1–3 on computation and go quickly to chapter 4, which focuses on data. These first chapters are important because all artificial intelligence (AI) is built on the same foundation of code, data, binary, and electrical impulses. Understanding what is real and what is imaginary in AI is crucial. Artificial superintelligences, like on the TV show *Person of Interest* or *Star Trek*, are imaginary. Yes, they're fun to imagine, and it can inspire wonderful creativity to think about the possibilities of robot domination and so on—but they aren't real. This book hews closely to the real mathematical, cognitive, and computational concepts that are in the actual academic discipline of artificial intelligence: knowledge representation and reasoning, logic, machine learning, natural language processing, search, planning, mechanics, and ethics.

In the first computational adventure (chapter 5), I investigate why, after two decades of education reform, schools still can't get students to pass standardized tests. It's not the students' or the teachers' fault. The problem is far bigger: the companies that create the most important state and local exams also publish textbooks that contain many of the answers, but low-income school districts can't afford to buy the books.

I discovered this thorny situation by building artificial intelligence software to enable my reporting. Robot reporters have been in the news in recent years because the Associated Press (AP) is using bots to write routine business and sports stories. My software wasn't inside a robot (it didn't need to be, although I'm not averse to the idea), nor did it write any stories for me (ditto). Instead, it was a brand-new application of old-school artificial intelligence that helped reveal some fascinating insights. One of the most surprising findings of this computational investigation was that, even in our high-tech world, the simplest solution—a book in the hands of a child—was quite effective. It made me wonder why we are spending so much money to put technology into classrooms when we already have a cheap, effective solution that works well.

The next chapter (chapter 6) is a whirlwind tour through the history of machines, specifically focused on Marvin Minsky—commonly known as the father of artificial intelligence—and the enormous role that 1960s counterculture played in developing the beliefs about the Internet that exist in 2017, the time this book was written. My goal here is to show you how the dreams and goals of specific individuals have shaped scientific knowledge, culture, business rhetoric, and even the legal framework of today's technology through deliberate choices. The reason we don't have national territories on the Internet, for example, is that many of the people who made the Internet believed they could make a new world beyond government— much like they tried (and failed) to make new worlds in communes.

In thinking about tech, it's important to keep another cultural touchstone in mind: Hollywood. A great deal of what people dream about making in tech is shaped by the images they see in movies, TV programs, and books. (Remember my childhood robot?) When computer scientists refer to *artificial intelligence*, we make a distinction between general AI and narrow AI. *General AI* is the Hollywood version. This is the kind of AI that would power the robot butler, might theoretically become sentient and take over the government, could result in a real-life Arnold Schwarzenegger as the Terminator, and all of the other dread possibilities. Most computer scientists have a thorough grounding in science fiction literature and movies, and we're almost always happy to talk through the hypothetical possibilities of general AI.

Inside the computer science community, people gave up on general AI in the 1990s.[3] General AI is now called *Good Old-Fashioned Artificial Intelligence* (GOFAI). *Narrow AI* is what we actually have. Narrow AI is purely mathematical. It's less exciting than GOFAI, but it works surprisingly well and we can do a variety of interesting things with it. However, the linguistic confusion is significant. Machine learning, a popular form of AI, is not GOFAI. Machine learning is narrow AI. The name is confusing. Even to me, the phrase *machine learning* still suggests there is a sentient being in the computer.

The important distinction is this: general AI is what we want, what we hope for, and what we imagine (minus the evil robot overlords of golden-age science fiction). Narrow AI is what we have. It's the difference between dreams and reality.

Next, in chapter 7, I define machine learning and demonstrate how to "do" machine learning by predicting which passengers survived the *Titanic* crash. This definition is necessary for understanding the fourth project (chapter 8), in which I ride in a self-driving car and explain why a self-driving school bus is guaranteed to crash. The first time I rode in a self-driving car was in 2007, and the computerized "driver" almost killed me in a Boeing parking lot. The technology has come a long way since then, but it still fundamentally doesn't work as well as a human brain. The cyborg future is not coming anytime soon. I look at our fantasies about technology replacing humans and explore why it's so hard to admit when technology isn't as effective as we want it to be.

Chapter 9 is a springboard for exploring why *popular* is not the same as *good* and how this confusion—which is perpetuated by machine-learning techniques—is potentially dangerous. Chapters 10 and 11 are also programming adventures, in which I start a pizza-calculating company on a cross-country hackathon bus trip (it's popular but not good) and try to repair the US campaign finance system by building AI software for the 2016 presidential election (it's good but not popular). In both cases, I build software that works—but it doesn't work as expected. Its demise is instructive.

My goal in this book is to empower people around technology. I want people to understand how computers work so that they don't have to be intimidated by software. We've all been in that position at one time or another. We've all felt helpless and frustrated in the face of a simple task that should be easy, but somehow isn't because of the technological interface. Even my students, who grew up being called digital natives, often find the digital world confusing, intimidating, and poorly designed.

When we rely exclusively on computation for answers to complex social issues, we are relying on artificial unintelligence. To be clear: it's the computer that's artificially unintelligent, not the person. The computer doesn't give a flying fig about what it does or what you do. It executes commands to the best of its abilities, then it waits for the next command. It has no sentience, and it has no soul.

People are always intelligent. However, smart and well-intentioned people act like technochauvinists when they are blind to the faults of computational decision making or they are excessively attached to the idea of using computers to the point at which they want to use computers for everything—including things for which the computer is not suited.

I think we can do better. Once we understand how computers work, we can begin to demand better quality in technology. We can demand systems that *truly* make things cheaper, faster, and better instead of putting up with systems that promise improvement but in fact make things unnecessarily complicated. We can learn to make better decisions about the downstream effects of technology so that we don't cause unintentional harm inside complex social systems. And we can feel empowered to say "no" to technology when it's not necessary so that we can live better, more connected lives and enjoy the many ways tech can and does enhance our world.

2 Hello, World

To understand what computers *don't* do, we need to start by understanding what computers do well and how they work. To do this, we'll write a simple computer program. Every time a programmer learns a new language, she does the same thing first: she writes a "Hello, world" program. If you study programming at coding boot camp or at Stanford or at community college or online, you'll likely be asked to write one. "Hello, world" is a reference to the first program in the iconic 1978 book *The C Programming Language* by Brian Kernighan and Dennis Ritchie, in which the reader learns how to create a program (using the C programming language) to print "Hello, world." Kernighan and Ritchie worked at Bell Labs, the think tank that is to modern computer science what Hershey is to chocolate. (AT&T Bell Labs was also kind enough to employ me for several years.) A huge number of innovations originated there, including the laser and the microwave and Unix (which Ritchie also helped develop, in addition to the C programming language). C got its name because it is the language that the Bell Labs crew invented after they wrote a language called "B." C++, a still-popular language, and its cousin C# are both descendants of C.

Because I like traditions, we'll start with "Hello, world." Please get a piece of paper and a writing utensil. Write "Hello, world" on the paper.

Congratulations! That was easy.

Behind the scenes, it was more complex. You formed an intention, gathered the necessary tools to carry out your intention, sent a message to your hand to form the letters, and used your other hand or some other parts of your body to steady the page while you wrote so that the physics of the situation worked. You instructed your body to follow a set of steps to achieve a specific goal.

Now, you need to get a computer to do the same thing.

Open your word-processing program—Microsoft Word or Notes or Pages or OpenOffice or whatever—and create a new document. In that document, type "Hello, world." Print it out if you like.

Congratulations again! You used a different tool to carry out the same task: intention, physics, and so on. You're on a roll.

The next challenge is to make the computer print "Hello, world" in a slightly different way. We're going to write a program that prints "Hello, world" to the screen. We're going to use a programming language called Python that comes installed on all Macs. (If you're not using a Mac, the process is slightly different; you'll need to check online for instructions.) On a Mac, open the Applications folder and then open the Utilities folder inside it. Inside Utilities, there's a program called *Terminal* (see figure 2.1). Open it.

Congratulations! You've just leveled up your computer skills. You're now close to the metal.

The metal means the computer hardware, the chips and transistors and wires and so on, that make up the physical substance of a computer. When

Figure 2.1
Terminal program in the Utilities folder.

you open the terminal program, you're giving yourself a window through the nicely designed graphical user interface (GUI) so that you can get closer to the metal. We're going to use the terminal to write a program in Python that will print "Hello, world" on the computer screen.

The terminal has a blinking cursor. This marks what's called the *command line*. The computer will interpret, quite literally and without any nuance, everything that you type in at the command line. In general, when you press Return/Enter, the computer will try to execute everything you just typed in. Now, try typing in the following:

```
python
```

You'll see something that looks like this:

```
Python 3.5.0 (default, Sep 22 2015, 12:32:59)
[GCC 4.2.1 Compatible Apple LLVM 7.0.0 (clang-700.0.72)] on
darwin
Type "help," "copyright," "credits" or "license" for more
information.
>>>
```

The triple-carat marks (>>>) tell you that you're in the Python interpreter, not the regular command-line interpreter. The regular command line uses a kind of programming language called a *shell programming language*. The Python interpreter uses the Python programming language instead. Just as there are different dialects in spoken language, so too are there many dialects of programming languages.

Type in the following and press Return/Enter:

```
print("Hello, world!")
```

Congratulations! You just wrote a computer program! How does it feel?

We just did the same thing three different ways. One was probably more pleasant than the others. One was probably faster and easier than the others. The decision about which one was easier and which one felt faster has to do with your individual experience. Here's the radical thing: *one was not better than the other.* Saying that it's better to do things with technology is just like saying it's better to write "Hello, world" in Python versus scrawling it on a piece of paper. There's no innate value to one versus the other; it's about how the individual experiences it and what the real-world consequences are. With "Hello, world," the stakes are very low.

Most programs are more complex than "Hello, world," but if you understand a simple program, you can scale up your understanding to more complex programs. Every program, from the most complex scientific computing to the latest social network, is made by people. All those people started programming by making "Hello, world." The way they build sophisticated programs is by starting with a simple building block (like "Hello, world") and incrementally adding to it to make the program more complex. Computer programs are not magical; they are made.

Let's say that I want to write a program that prints "Hello, world" ten times. I could repeat the same line many times:

```
print("Hello, world!")
print("Hello, world!")
```

Ugh. Nope, not going to do that. I'm already bored. Pressing Ctrl+P to paste eight more times would require far too many keystrokes. (To think like a computer programmer, it helps to be lazy.) Many programmers think typing is boring and tedious, so they try to do as little of it as possible. Instead of retyping, or copying and pasting, the line, I'm going to write a loop to instruct the computer to repeat the instruction ten times.

```
x=1
while x<=10:
    print("Hello, world!\n")
    x+=1
```

That's way more fun! Now the computer will do all the work for me! Wait—what just happened?

I set the value of x to 1 and created a WHILE loop that will run until it reaches the stop condition, x>10. On the first time through the loop, x=1. The program prints "Hello, world!" followed by a carriage return, or end of line character, which is indicated by \n (pronounced backslash-n). A backslash is a special character in Python. The Python interpreter is programmed to "know" that when it reads that special character, it should do something special with the text that happens immediately afterward. In this case, I am telling the computer to print a carriage return. It would be a pain to start from scratch every time and program each dumb hunk of metal to perform the same underlying functions, like read text and convert it to binary, or to carry out certain tasks according to the conventions of the syntax of our chosen programming language. Nothing would ever get done! Therefore, all computers come with some built-in functions and with the ability to add functions. I use the term *know* because it's convenient, but please remember that the computer doesn't "know" the way that a sentient being "knows." There is no consciousness inside a computer; there's only a collection of functions running silently, simultaneously, and beautifully.

In the next line, x+=1, I am incrementing x by one. I think this stylistic convention is particularly elegant. In programming, you used to have to write x=x+1 every time you wanted to increment a variable to make it through the next round of a loop. The Python stylists thought this was boring, and they wrote a shortcut. Writing x+=1 is the same as writing x=x+1. The shortcut is taken from C, where a variable can be incremented with the notation x++ or ++x. There are similar shortcuts in almost every programming language because programmers do a *lot* of incrementing by one.

After one increment, x=2, and the computer hits the bottom of the loop. The indentation of the lines under the *while* statement mean that these lines are part of the loop. When it reaches the end of the loop, the computer goes back to the top of the loop—the *while* line—and evaluates the condition again: is x<=10? Yes. Therefore, the computer goes through the instructions again and prints "Hello, world!\n" which appears on the screen like this:

```
Hello, world!
```

Then, it increments x again. Now, x=3. The computer returns to the top of the loop again, and again—until x=11. When x=11, the stop condition is met, so the loop ends. Here's another way of thinking about it:

```
IF: x<=10
THEN: DO_THE_INSTRUCTIONS_INSIDE_THE_LOOP
ELSE: PROCEED_TO_THE_NEXT_STEP.
```

Each routine (or subroutine) is a small step. If you assemble a lot of small steps together, you can do very big things. Computer programmers get very good at looking at a task, breaking the task down into small parts, and programming the computer to take care of each of the small parts. Then, you put the parts together and tinker with them a bit to make them work with each other, and soon you have a working computer program. Today's programs are modular, meaning that one programmer can build the first module, another programmer can build the second module, and both modules will be able to work together if they're hooked up the right way.

Now that we've written a program, let's talk about data. Data can be the input or the output of a program. We generate data, meaning information points or units of information, about the world in a variety of ways. The National Weather Service gathers data on the high and low temperatures in thousands of American locales each day. A pedometer can track the number of steps you take in a day, yielding a pattern of steps taken in a day, a week, or a year. A kindergarten teacher I know has his students tally up the number of pockets in his classroom on Mondays. Data can show us the number of people who bought a particular hat; it can show us how many endangered white rhinos are left in the wild; it can show us the rate at which the polar ice caps are melting. Data is fascinating. It gives us insights. It allows us to learn about the world and to grapple with concepts that are beyond our current understanding. (Although if you're old enough to read this book, hopefully you've already come to grips with the idea of other people's pockets.)

Although the data may be generated in different ways, there's one thing all the preceding examples have in common: all of the data is generated by people. This is true of *all* data. Ultimately, data always comes down to people counting things. If we don't think too hard about it, we might imagine that data springs into the world fully formed from the head of Zeus. We assume that because there is data, the data must be true. Note the first principle of this book: *data is socially constructed*. Please let go of any notion that data is made by anything except people.

"What about computer data?" a savvy kindergarten pocket-data collector might ask. That's a very good question. Data generated by computers is ultimately socially constructed because people make computers. Math is a system of symbols entirely created by people. Computers are machines that compute: they perform millions of mathematical calculations. Computers are not built according to any kind of absolute universal or natural principles; they are machines that result from millions of small, intentional design decisions made by people who work in specific organizational contexts. Our understanding of data, and the computers that generate and process data, must be informed by an understanding of the social and technical context that allows people to make the computers that make the data.

One way to understand what comes out of computers is to understand what goes *into* computers. There are certain physical realities to the computer. Most computers are protected by a hard case, and inside the case is a bunch of circuit boards and stuff. Let me be more specific about this *stuff*. The important parts are the power source, the connection to the screen, the transistors, the built-in memory, and the writeable memory. All these things fall under the category of *hardware*. Hardware is physical; software is anything that runs on top of the hardware.

I first learned about the physical reality of a computer in high school in the 1990s. I was in a special engineering program for kids that was sponsored by Lockheed Martin. There was a Lockheed plant in my small New Jersey town. The building was shaped like a battleship and was surrounded by miles of unused farmland. The rumor back then was that the plant manufactured nuclear weapons, and that under the amber waves of grain were missile silos that would rise and shoot nuclear missiles in the case of attack by the Soviet Union. This was just before the end of the Cold War era, and everyone had seen the terrifying TV movie *The Day After* about the aftermath of a nuclear apocalypse, so we regularly had conversations about where the US missiles were, where the Soviets' missiles would land, and what we would do afterward. A few times a month, I took a school bus to the Lockheed plant to meet up with a handful of other teenagers from local schools and learn about engineering.

People sometimes say that a computer is like a brain. It isn't. If you take a piece out of a brain, the brain will reroute pathways to compensate. Think about the traumatic brain injury suffered by Arizona Congresswoman Gabby Giffords in 2011. Giffords was holding a meeting with constituents in the parking lot of a Safeway grocery store when a lone gunman, Jared Lee Loughner, shot her in the head at point-blank range. Loughner next shot blindly around the parking lot, killing six people and wounding eighteen. He had been stalking Giffords.

Giffords's intern, Daniel Hernandez Jr., held her upright and applied pressure to the wound while bullets flew through the parking lot. Eventually, bystanders subdued Loughner and police and emergency services arrived. Giffords was in critical condition. Doctors performed emergency brain surgery and then put her into a medically induced coma to allow her brain to heal. Four days after the attack, Giffords opened her eyes. She couldn't speak, she could barely see—but she was alive.

Giffords courageously faced the long road to recovery. With intensive therapy, she relearned how to speak. Like most people who suffer this kind of traumatic brain injury, Giffords's voice was very different than it was before the attack. Her new voice was slower, and her speech sounded labored. Speaking left her tired. Her brain created new pathways that were different than the old, missing pathways. This is one of the amazing things that a brain can do: it can, under very specific conditions and in very specific ways, repair itself.

A computer can't do this. If you take a piece out of a computer, it simply won't work. Everything stored in computer memory has a physical address. The working draft of this book is stored in a particular spot on my computer's hard drive. If that spot was erased, I would lose all these carefully crafted pages. It would be bad; I might have a small breakdown and miss my deadline. However, the ideas would still exist in my brain, so I could recreate the text if necessary. A brain is more flexible and adaptable than a hard drive.

This was one of the many useful things I learned at Lockheed. I also discovered that at tech companies, there are always plenty of slightly outdated spare parts lying around because people upgrade their computers or leave the company. Each teenager in the program was given a case for an Apple II computer, a circuit board, some memory chips, some brightly colored ribbon cables, and miscellaneous other parts scavenged from various offices in the (possibly nuclear) plant. We plugged these components in, and our teacher explained what each part did. The cases were dirty and the keyboards were slightly sticky and all the circuit boards were dusty, but we didn't care. We were building our own computers, and it was fun. After we built our computers, we learned to program them using a simple programming language called BASIC. At the end of the semester, we got to keep the computers.

I tell this story because it's important to think of a computer as an object that can be and is constructed by human hands. Often, the students who show up in my programming for journalists classes are intimidated by technology. They worry that they are going to break the computer or make some kind of catastrophic misstep. "The only way you can break the computer is with a hammer," I tell them. They rarely believe me at first. By the end of the semester, they are more confident. Even if they break something, they have faith that they can fix it or figure it out. This confidence is key in technological literacy.

You are not in my classroom, so I can't hand you a computer, but I encourage you to take apart an old one. You may have one lying around; otherwise, old computers are often available at thrift stores for not very much money. You might ask around at an office; usually, the system administrator or web person will have some old technology lying around as decoration, or because they haven't gotten around to recycling it yet. A desktop computer is the easiest to use for this activity.

Take the computer apart. You'll probably need a very small screwdriver if you are dismantling a laptop. The interior of the desktop computer probably looks something like the image in figure 2.2.

Look at the parts, how they are put together. Follow the wires from the inputs (USB port, video port, speaker port, etc.) and see where they connect. Touch the rectangular blobs that seem cemented on the circuit board. Find the microprocessor chips: these are the pieces that probably say "Intel" and are the key to this whole endeavor. They are important. Find the plug that connects the computer hardware to the monitor. It's probably connected to an extremely strong, flexible, plasticky ribbon. This carries information about graphics to the screen, then the screen displays the graphics specified in the code.

When you wrote your Python program, you typed on the keyboard. That information was carried into the computer body from the keyboard, then was interpreted character by character. Then, the computer sent out an instruction from the body to another part of the machine—the monitor—telling it to print the text "Hello, world." This cycle happens over and over again, with simple or complex instructions.

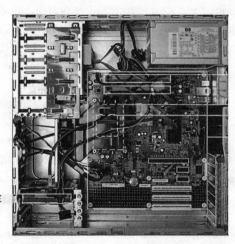

POWER SUPPLY

RAM
(RANDOM ACCESS MEMORY)

HEAT SINK
CPU
(CENTRAL PROCESSING UNIT)
MOTHERBOARD

EXPANSION SLOTS

HARD DRIVE

Figure 2.2
The innards of a desktop computer.

Dismantling a computer is a great activity to do with a kid. I once took apart a laptop with my son when he was in elementary school. I wanted to recycle a couple of laptops, and I was pulling out the hard drives to smash them with a hammer before dropping them at the recycler. (I discovered at some point that smashing a hard drive is easier, and often more satisfying, than erasing it.) I asked my son if he wanted to help me take the hard drive out of the computer. "Are you kidding? I want to take the whole thing apart," he said. So, we spent an enjoyable hour or two disassembling the two laptops on the kitchen counter.

In my university class, we play with hardware and then move on to talking about software—including "Hello, world." Software is everything that runs on top of the hardware. It's what allows you to write an instruction on the keyboard and have the machine act on the instruction. It's what allows the "Hello, world" program to run. Behind the scenes, the text you write is being compiled into instructions that the machine can follow. Hardware is physical; software is everything else. *Computer programming* and *writing software* usually are the same thing.

I'm not going to lie: programming is math. If anyone tries to convince you that it isn't, or that you can really learn programming without doing math, they're probably trying to sell you something.

The good news is, the math that you need for introductory programming is the math you learn around fourth or fifth grade. You'll need to have mastered addition, subtraction, multiplication, division, fractions, percentages, and remainders. You'll need basic geometry like area, perimeter, radius, circumference. You'll need to know basic graphing terms like x, y, and z axes. Finally, you'll need to know the basics of functions—that to turn 2 into 22, we perform a mathematical function on it.

If you have a major math phobia, you probably want to stop reading now. That's OK! There's a lot of rhetoric out there that suggests everyone should learn to code. I don't agree with this. If you really can't do math, coding will probably make you miserable. However, if you're confident that you can calculate the tip at a restaurant, and you can do everyday things like estimate how big a rug to get for your living room, you'll be fine.

To get beyond introductory programming to intermediate programming requires knowing linear algebra, some geometry, and some calculus. However, many people do just fine in their careers with "only" introductory programming skills. Programming can be both an art and a craft. For programming as a craft, you can apprentice and learn and earn a decent living. Programming as an art requires craftsmanship plus training in advanced mathematics. This book assumes you're primarily interested in craft.

There are technical ways to describe how software and hardware work together. For the moment, I'm going to use a metaphor instead. Understanding the layers of a computer is like understanding the layers of a turkey club sandwich (figure 2.3).

The turkey club is a familiar sight. It has lots of different parts, but they all work well together and result in a delicious sandwich. Just like you build a turkey club in a specific order to achieve a certain effect, a computer runs in a specific order.

Building a turkey club starts with the base layer of bread. That's like the hardware in a computer. The hardware doesn't "know" anything—it just knows how to deal with binary data, 0s and 1s. By *deal with*, I mean

Figure 2.3
A turkey club sandwich.

calculate. Remember that everything a computer can do comes down to math.

On top of the hardware is a layer that allows you to translate words into binary (0s and 1s). Let's call this the *machine-language layer*. It's like the layer of turkey that comes next in the club sandwich. Machine language translates symbols into binary so that the computer can perform calculations. Those symbols are the words and numbers that we humans use to communicate meaning to each other. It's a constructed system. The dialect you use to "speak" machine language is called *assembly language*. It assembles symbols into machine code.

Assembly language is difficult. Here's a sample of an assembly language program to write "Hello, world" ten times, which I copied from a post on a developers' site called Stack Overflow:

```
org
      xor ax, ax
      mov ds, ax
      mov si, msg
boot_loop:lodsb
      or al, al
      jz go_flag
      mov ah, 0x0E
      int 0x10
      jmp boot_loop
go_flag:
      jmp go_flag
msg db 'hello world', 13, 10, 0
      times 510-($-$$) db 0
      db 0x55
      db 0xAA
```

Assembly language is not easy to read or write. Very few people want to spend their days in this language. To make it easier for humans to communicate instructions, we put something on top of the machine-language layer. This is called an *operating system*. On my Mac, the operating system is Linux, which is named after its creator, Linus Torvalds. Linux is based on Unix, the operating system developed by Ritchie of "Hello, world" fame. You probably know operating systems well, even if you don't know what they're called. Part of the personal computer revolution of the 1980s was the triumph of operating systems, which run on top of the machine-language layer and are far easier to interact with if you're a human.

At this point, you have a perfectly serviceable (if plain) computer. You can run all kinds of exciting, interesting programs just using Linux. However, Linux is primarily text-based, and it's not intuitive—so, on the Mac, there's another operating system, OSX, the recognizable Mac interface. It's called a graphical user interface (GUI). The GUI was one of Steve Jobs's great innovations: he realized that using the text-based interface was difficult, so he popularized the practice of putting pictures (icons) on top of the text and using the mouse as a way of navigating among the pictures. Jobs got the idea of the desktop GUI and the mouse from Alan Kay's team at Xerox PARC, another research lab, which released a computer with a GUI and mouse in 1973. Although we like to credit individuals for technological innovations, rarely is it the case that a lone inventor created any modern computational innovations. When you look closely, there's always a logical predecessor and a team of people who worked on the idea for months or years. Jobs paid for a tour of Xerox PARC, saw the idea of a GUI, and licensed it. The Xerox PARC mouse-and-GUI computer was a derivative of an earlier idea, the oN-Line System (NLS), demonstrated by Doug Engelbart in the "mother of all demos" at the 1968 Association for Computing Machinery conference. We'll look at this intricate history in chapter 6.

The next layer to think about is another software layer: a program that runs on top of an operating system. A web browser (like Safari or Firefox or Chrome or Internet Explorer) is a program that allows you to view web pages. Microsoft Word is a word processing program. Desktop video games like Minecraft are also programs. These programs are all designed to take advantage of certain underlying features of the different operating systems. That's why you can't just run a Windows program on a Mac (unless you use another software program—an emulator—to help you). These programs are designed to seem very easy to use, but underneath they're highly precise.

Let's add some complexity. Imagine that you're a journalist who writes a weekly online column about cats. You use a software program to compose your column. Most journalists compose in a word processing program like Microsoft Word or Google Docs. Either of these programs can run either locally or in the cloud. *Locally* means that the program is running on the hardware on your computer. *In the cloud* means that the program is running on someone else's computer. The cloud is a wonderful metaphor, but practically speaking, *the cloud* just means "a different computer, probably located with thousands of other computers in a large warehouse in the tristate area." The content you create is the truly unique part that comes from your imagination: your elegant, pithy, lovingly crafted story about cats riding Roomba vacuum cleaners or whatever. To the computer, every story is the same, just a collection of 0s and 1s stored on a hard drive somewhere.

After you compose your story, you put it into a content-management system (CMS) so that it can be seen by your editor and eventually by your audience. A CMS is an essential piece of software for the modern media organization. Media organizations handle hundreds of stories each day, every day. Each story is due at a different time of day; each story is in a different state of editing (or disarray) at any given time; each story has a different headline for print and for online; each story has a different excerpt to be used on each social media platform; each story has images or video or data visualizations or code associated with it; each story is created by a person who needs to be complimented or paid or managed; and all this goes on 24 hours a day, 365 days a year. The scale is vast. I can't stress this enough. It would be foolish to try to manage this type of endeavor without software. The CMS is a tool for managing all the stories and images and so forth that the media organization publishes in print or online.

The CMS also allows the media organization to apply a uniform design template to each story so that the stories all look similar. This is good for branding, but it's also practical. If every single story had to be individually designed for digital presentation, it would take forever to publish anything. Instead, the CMS imposes a standardized design template on top of the raw text that you, the reporter, type into the CMS.

Consider the process of deciding what parts of the design template you will use in your story to decorate it. Will you use pull quotes? Will you include hyperlinks? Will you embed social media posts by people you quote in the story? These are all small design decisions that will affect the reader's experience of your story.

Finally, the story needs to go out into the world. A web server, another piece of software, is used to take the story from the CMS to a person who wants to read the story. The reader accesses the story via a web browser like Chrome or Safari. The web browser is called a *client*. The web server serves the story (which the CMS converts to an HTML page) to the client. The client-server model, the endless sending and receiving of information, is how the web works. The terms *client* and *server* come from restaurants. One way to understand the client-server model is to think about a human server at a restaurant, who distributes food to human clients of the restaurant.

This is the underlying process (more or less) every time you access something on the web. There are many steps and thus many opportunities for things to go wrong. Really, it's quite impressive that things don't go wrong more often.

Every time you use a computer, you are using this complex set of layers. There is no magic to it, although the results can seem amazing. Understanding the technical realities is important because it allows you to anticipate how, why, and where things will go wrong in a computerized scenario. Even if you feel like the computer is talking to you, or you feel like you are having an interaction with a computer, what you are really doing is having an interaction with a program written by a human being with thoughts, feelings, biases, and background.

This often works out beautifully. It is straight-up fun to interact with Eliza, the 1966 text interaction bot that responds to questions in the manner of a Rogerian psychotherapist. To this day, there are bots on Twitter that respond to users with the patterns pioneered by the Eliza software. A simple Internet search will turn up many examples of Eliza code.[1] Eliza's canned responses are based on the user's input. The replies include the following:

```
Don't you believe that I can _____?
Perhaps you would like to be able to _____.
You want me to be able to _____.
Perhaps you don't want to _____.
Tell me more about such feelings.
What answer would please you the most?
What do you think?
What is it you really want to know?
Why can't you _____?
Don't you know?
```

Try to build an Eliza bot, and the limitations of the form quickly become apparent. Can you build a set of responses that work in any situation? No way. You can think of responses that would suit most situations, but not all. There will always be limitations to what a computer can say in response to a human, because there will always be limits to the imagination of the human computer programmer. Even crowdsourcing will not be adequate, because there will never be enough people to predict every situation that has ever arisen or will ever arise in the future. The world changes; so do conversational styles. Even Rogerian therapy is no longer considered the latest and greatest interaction style; cognitive behavioral therapy is far more in vogue right now.

Trying to predict every possible response for a bot is doomed in part because we can't get away from unforeseen events. I'm reminded of the time that I found out a friend committed suicide by jumping in front of a New York City subway train. I didn't know this was coming, and I didn't know what to do once I heard. For a while, everything seemed to stop.

Eventually, the shock passed and I began to mourn. But until it happened, I didn't have any way to predict that this particular tragedy would be something that I'd have to assimilate. We're all the same in this regard. Programmers are no better than anyone else at anticipating unexpected, terrible situations. Social groups tend to have a collective blind spot when it comes to imagining the worst. It's a kind of cognitive bias that sociologist Karen A. Cerulo calls "positive asymmetry" in her book *Never Saw It Coming: Cultural Challenges to Envisioning the Worst*. Positive asymmetry is a "tendency to emphasize only the best or most positive cases," she writes. Cultures tend to reward those who focus on the positive and shun or punish those who bring up the downside. The programmer who brings up the potential new audience for a product gets more attention than the programmer who points out that the new product will likely be used for harassment or fraud.[2]

Eliza's responses reflect its designer's basically playful outlook. Looking at Eliza's responses, it's easy to see how voice assistants like Apple's Siri are programmed. The original Eliza had a few dozen responses; Siri includes many, many responses crafted by many, many people. Siri can do a lot: it can send messages, place phone calls, update a calendar with appointments, or set an alarm. It can be fun to stump Siri. Little kids take especial delight in testing the outer limits of what Siri will say. However, Siri and the other voice assistants are limited in their verbal responses by the collective imagination (and positive asymmetry) of their programmers. A team at the Stanford School of Medicine tested the various voice assistants to see whether the assistants recognized a health crisis, responded with respectful language, and referred the person to an appropriate resource. The programs responded "inconsistently and incompletely," the authors wrote in *JAMA Internal Medicine* in 2016. "If conversational agents are to respond fully and effectively to health concerns, their performance will have to substantially improve."[3]

Technochauvinists like to believe that computers do a better job than people at most tasks. Because the computer operates based on mathematical logic, they think that this logic translates well to the offline world. They are right about one thing: when it comes to calculating, computers do a far better job than people alone. Anyone who has ever graded a student math paper will happily admit that. But there are limits to what a computer can do in certain situations.

Consider the tacocopter, a fanciful idea that had a moment of online popularity. It sounds delightful: a quadcopter drone that delivers a hot, tasty bag full of tacos right to your door! However, when you think about the hardware and software, the flaws in the idea become apparent. A drone is basically a remote-controlled helicopter with a computer and a camera. What happens when it rains? Electrical things don't do well in rain, snow, or fog. My cable television service always malfunctions in a rainstorm, and a wireless drone is far more fragile. Is the tacocopter supposed to come to the window? The front door? How will it push the button in the elevator, or open a stairway door, or push an intercom bell? These are all mundane tasks that are easy for humans, but insanely difficult for computers. How might a tacocopter be co-opted to deliver other, less nutritious and legal substances? What would happen when it inevitably gets shot out of the sky by a freaked-out homeowner with a gun? Only a technochauvinist would imagine that a tacocopter is better than the human-based system that we have now.

If you ask Siri if tacocopters are a good idea, she will look up that phrase for you online. What you'll get are a bunch of news articles about the tacocopter, including one from *Wired* magazine (more on that publication and one of its founders, Stewart Brand, in chapter 6) that debunks the concept more fully than I have done here. The founder admits it's logistically impossible, not least because of FAA regulations on the commercial use of unmanned aerial vehicles. But, she claims, keeping the vision of the idea alive is still important. "Like what cyberpunk did for the internet," she says. "Mull the possibilities, give people things to think about."[4]

What seems to be missing here is a more complete vision of what a world with functioning tacocopters would be like. What would it mean to design buildings and urban environments to enable drones instead of humans? How would our access to light and air change if windows became docking stations for food-delivery vehicles? What might be the social costs of eradicating even that most mundane and insignificant of interactions—a bag of food being passed from one human hand to another? Do we really want to say "Hello, world" to that reality?

3 Hello, AI

We've covered hardware, software, and programming. It's time to move on to a more advanced programming topic: artificial intelligence. To most people, the phrase *artificial intelligence* suggests something cinematic—maybe Commander Data, the lifelike cyborg from *Star Trek: The Next Generation*; perhaps Hal 9000 from *2001: A Space Odyssey*; or Samantha, the AI system from the movie *Her*; or Jarvis, the AI majordomo that helps Iron Man in the Marvel comics and movies. Regardless, here's what's important to remember: *those are imaginary*. It's easy to confuse what we imagine and what is real—especially when we want something very badly. Many people want AI to be real. This usually takes the form of wanting a robot butler to attend to your every need. (I will confess to having had many late-night undergraduate conversations about the practical and ethical considerations of having a robot butler.) A disproportionate number of the people who make tech fall into the camp of desperately wanting Hollywood robots to be real. When Facebook's Mark Zuckerberg built an AI-based home automation system, he named it Jarvis.

One excellent illustration of the confusion between real and imaginary AI happened to me at the NYC Media Lab's annual symposium, a kind of science fair for grownups. I was giving a demo of an AI system I built. I had a table with a monitor and a laptop hooked up to show my demo; three feet away was another table with another demo by an art school undergraduate who had created a data visualization. Things became boring when the crowd died down, so we got to chatting.

"What's your project?" he asked.

"It's an artificial intelligence tool to help journalists quickly and efficiently uncover new story ideas in campaign finance data," I said.

"Wow, AI," he said. "Is it a *real* AI?"

"Of course," I said. I was a little offended. I thought: *Why would I spend my day demonstrating software at this table if I hadn't made something that worked?*

The student came over to my table and started looking closely at the laptop hooked up to the monitor. "How does it work?" he asked. I gave him the three-sentence explanation (you'll read the longer explanation in chapter 11). He looked confused and a little disappointed.

"So, it's not real AI?" he asked.

"Oh, it's real," I said. "And it's spectacular. But you know, don't you, that there's no simulated person inside the machine? Nothing like that exists. It's computationally impossible."

His face fell. "I thought that's what AI meant," he said. "I heard about IBM Watson, and the computer that beat the champion at Go, and self-driving cars. I thought they invented real AI." He looked depressed. I realized he'd been looking at the laptop because he thought there was something in there—a "real" ghost in the machine. I felt terrible for having burst his bubble, so I steered the conversation toward a neutral topic—an upcoming Star Wars movie—to cheer him up.

This interaction stuck with me because it helps me remember the difference between how computer scientists think about AI and how members of the public—including highly informed undergraduates working on tech—think about AI.

General AI is the Hollywood kind of AI. General AI is anything to do with sentient robots (who may or may not want to take over the world), consciousness inside computers, eternal life, or machines that "think" like humans. *Narrow AI* is different: it's a mathematical method for prediction. There's a lot of confusion between the two, even among people who make technological systems. Again, general AI is what some people want, and narrow AI is what we have.

One way to understand narrow AI is this: narrow AI can give you the most likely answer to any question that can be answered with a number. It involves quantitative prediction. Narrow AI is statistics on steroids.

Narrow AI works by analyzing an existing dataset, identifying patterns and probabilities in that dataset, and codifying these patterns and probabilities into a computational construct called a *model*. The model is a kind of black box that we can feed data into and get an answer out of. We can take the model and run new data through it to get a numerical answer that predicts something: how likely it is that a squiggle on a page is the letter *A*; how likely it is that a given customer will pay back the mortgage money a bank loans to him; which is the best next move to make in a game of tic-tac-toe, checkers, or chess. Machine learning, deep learning, neural networks, and predictive analytics are some of the narrow AI concepts that are currently popular. For every AI system that exists today, there is a logical explanation for how it works. Understanding the computational logic can demystify AI, just like dismantling a computer helps to demystify hardware.

AI is tied up with games—not because there's anything innate about the connection between games and intelligence, but because computer scientists tend to like certain kinds of games and puzzles. Chess, for example, is quite popular in their crowd, as are strategy games like Go and backgammon. A quick look at the Wikipedia pages for prominent venture capitalists and tech titans reveals that most of them were childhood Dungeons & Dragons enthusiasts.

Ever since Alan Turing's 1950 paper that proposed the *Turing test* for machines that think, computer scientists have used chess as a marker for "intelligence" in machines. Half a century has been spent trying to make a machine that could beat a human chess master. Finally, IBM's Deep Blue defeated chess champion Garry Kasparov in 1997. AlphaGo, the AI program that won three of three games against Go world champion Ke Jie in 2017, is often cited as an example of a program that proves general AI is just a few years in the future. Looking closely at the program and its cultural context reveals a different story, however.

AlphaGo is a human-constructed program running on top of hardware, just like the "Hello, world" program you wrote in chapter two. Its developers explain how it works in a 2016 paper published in *Nature*, the international journal of science.[1] The opening lines of the paper read: "All games of perfect information have an optimal value function, $v*(s)$, which determines the outcome of the game, from every board position or state s, under perfect play by all players. These games may be solved by recursively computing the optimal value function in a search tree containing approximately b^d possible sequences of moves, where b is the game's breadth (number of legal moves per position) and d is its depth (game length)." This is perfectly clear to someone who has years of high-level mathematical training, but many of us would prefer a plainer-language explanation.

To understand AlphaGo, it helps to start by thinking about tic-tac-toe, a game that most children have mastered. If you go first in tic-tac-toe and choose the space in the middle of the nine-square grid, you can always play to a win or a draw. Going first gives you an advantage: you will have five moves to your opponent's four. Most kids grasp this intuitively and insist on going first when playing with an indulgent older opponent.

It's also relatively easy to write a computer program to play tic-tac-toe against a human opponent. The first one was written in 1952. There's an *algorithm*, a set of rules or steps, that you can deploy so that the computer always plays to a win or a draw. Like "Hello, world," building a tic-tac-toe game is a common exercise in introductory computing classes.

Go is far more sophisticated than tic-tac-toe, but it's also a game played on a grid. Each Go player receives a pile of either black or white stones. Beginners play on a grid made of nine vertical and nine horizontal lines; advanced players use a nineteen-by-nineteen grid. Black goes first and places a black stone at an intersection of two lines. White then places her stone at a different intersection. The players alternate turns, with the goal of "capturing" the opponent's stones by surrounding a stone with the opposite color.

People have been playing Go for three thousand years. Computer scientists and Go aficionados have been studying patterns in the game since at least 1965. The first computerized Go program was written in 1968. There's an entire subfield of computer science research devoted to Go, called (unsurprisingly) *Computer Go*.

For years, Computer Go players and researchers have been amassing records of games. A game record looks like this:

```
(;GM[1]
FF[4]
SZ[19]
PW[Sadavir]
WR[7d]
PB[tzbk]
BR[6d]
DT[2017-05-01]
PC[The KGS Go Server at http://www.gokgs.com/]
KM[0.50]
RE[B+Resign]
RU[Japanese]
```

```
CA[UTF-8]
ST[2]
AP[CGoban:3]
TM[300]
OT[3x30 byo-yomi]
;B[qd];W[dc];B[eq];W[pp];B[de];W[ce];B[dd];W[cd];B[ec];W[cc];B
[df];W[cg];B[kc];W[pg];B[pj];W[oe];B[oc];W[qm];B[of];W[pf];B[p
e];W[og];B[nf];W[ng];B[nj];W[lg];B[mf];W[lf];B[mg];W[mh];B[me]
;W[li];B[kh];W[lh];B[om];W[lk];B[qo];W[po];B[qn];W[pn];B[pm];W
[ql];B[rq];W[qq];B[rm];W[rl];B[rn];W[rj];B[qr];W[pr];B[rr];W[m
n];B[qi];W[rh];B[no];W[on];B[nn];W[nm];B[nl];W[mm];B[ol];W[mp]
;B[ml];W[ll];B[np];W[nq];B[mo];W[mq];B[lo];W[kn];B[ri];W[si];B
[qj];W[qk];B[kq];W[kp];B[ko];W[jp];B[lp];W[lq];B[jq];W[jo];B[j
n];W[in];B[lm];W[jm];B[ln];W[hq];B[qh];W[rg];B[nh];W[re];B[rd]
;W[qe];B[pd];W[le];B[md])
```

The text may look like gobbledygook to a human, but it's highly structured so that a machine can process it easily. The structure is called *smart games format* (SGF). The text shows who played the game, where, what each of the moves were, and how the game was resolved.

The large text area shows all the moves. Columns in the Go grid are labeled in alphabetical order from left to right, and rows are labeled from top to bottom. In this game, Black (B) went first and placed a stone at the intersection of column q and row d. This is shown as ;B[qd]. Then, the text ;W[dc] shows that White (W) placed a stone at the intersection of column d and column c. Each subsequent move is listed in this format. The resolution (RE) of the game is shown in the text RE[B+Resign], which means that Black resigned the game.

The AlphaGo designers amassed a massive dataset of thirty million SGF game files. The dataset wasn't randomly generated; those thirty million games were actual games played by actual people (and some computers). Whenever amateurs or professionals played Go on one of many online sites, that data was saved. It's not hard to create a Go video game; many versions of instructions and free code are posted online. All video games *can* save game data, of course. Some do; some don't. Some save your game data and use it for creating reports for the game company. The people who ran various online Go sites decided to publish their saved game data online in huge batches. Eventually, these batches were pooled, resulting in the thirty million games collected by the AlphaGo team.

The programmers used the thirty million games to "train" the model that they named *AlphaGo*. What you must remember is that people who play Go professionally spend *ages* playing Computer Go. It's how they train. Therefore, the thirty million games recorded included data from the world's greatest Go players. Millions of hours of human labor went into creating the training data—yet most versions of the AlphaGo story focus on the magic of the algorithms, not the humans who invisibly and over the course of years worked (without compensation) to create the training data.

The developers programmed AlphaGo to use a method called *Monte Carlo search* to pick a set of moves from the thirty million games that would most likely lead to a win. Then, they instructed it to use an algorithm to select the next move from the set. They also instructed it to use a different algorithm that calculated the probability of a win for each possible move in the set. The calculations happened on a scale that the human mind can barely imagine. There are 10^{170} possible board configurations in Go. By layering a variety of computational methods and always choosing the move with the greatest probability of success, the designers created a program that defeated the world's greatest Go players.

Is AlphaGo smart? Its designers certainly are. They solved a math problem that was so hard that it took decades of great minds to work on it. One of the amazing things about math is that it allows you to see underlying patterns in how the world works. Many, many things operate according to mathematical patterns: crystals grow in regular patterns, and cicadas hibernate underground for years and emerge when soil temperature conditions are just right, to name just two. AlphaGo is a remarkable mathematical achievement that was made possible by equally remarkable advances in computing hardware and software. AlphaGo's team of designers deserves praise for this outstanding technical achievement.

AlphaGo is not an intelligent machine, however. It has no consciousness. It does only one thing: plays a computer game. It contains data from thirty million games played by amateurs and by the world's most talented players. On some level, AlphaGo is supremely dumb. It uses brute force and the combined effort of many, many humans to defeat a single Go master. The program and its underlying computational methods will likely be deployed for other useful tasks involving massive number-crunching, and that's good for the world—but not everything in the world is a calculation.

Once we get past the mathematical and physical reality of a program like AlphaGo, we're in the realm of philosophy and future speculation. Those are very different intellectual landscapes. There are futurists who *want* AlphaGo to signify the beginning of an era in which people and machines become fused. Wanting something doesn't make it true, however.

Philosophically, there are lots of interesting questions to discuss centering on the difference between calculation and consciousness. Most people are familiar with the Turing test. Despite what the name suggests, the Turing test is not a quiz that a computer can pass to be considered intelligent. In his paper, Turing proposed a thought experiment about talking to a machine. He rejected the question "Can machines think?" as absurd and claimed it was best answered by an opinion poll. (Turing was a bit of a snob about math. Like many mathematicians then and a smaller number now, he believed in the superiority of mathematics to other intellectual pursuits.) Instead, Turing proposed an "imitation game" played by a man (A), a woman (B), and an interrogator (C). C sits in a room alone and submits typewritten queries to A and B. Turing writes: "The object of the game for the interrogator is to determine which of the other two is the man and which is the woman. He knows them by labels X and Y, and at the end of the game he says either 'X is A and Y is B' or 'X is B and Y is A.'"[2]

Turing then breaks down the kind of kind of questions the interrogator is allowed to ask. One is about hair length. A, the man, wants the interrogator to make the wrong assumptions and is willing to lie. B, the woman, wants to help the interrogator and can tell him or her that she is the woman—but A can lie and say that as well. Their answers are written down, so that the quality and tone of voice cannot provide clues. Turing writes: "We now ask the question, 'What will happen when a machine takes the part of A in this game?' Will the interrogator decide wrongly as often when the game is played like this as he does when the game is played between a man and a woman? These questions replace our original, 'Can machines think?'"

If the questioner can't tell the difference between a response provided by a human or the response provided by a machine, the computer is said to be *thinking*. For many years, this was considered foundational in computing. A vast amount of ink has been spilled trying to respond to Turing's ideas in this paper and to make a machine that can perform to Turing's specifications. However, undergirding the entire thought experiment is a philosophical and cultural misnomer that throws the entirety into question, and that is gender. Turing's specifications do not conform to what we now understand about gender. Gender is not a binary, but a continuum. Hair length is no longer a signifier of male or female identity; anyone can rock a short haircut. Moreover, as Turing writes, "The object of the game for the third player (B) is to help the interrogator." A game to determine "intelligence," in which the woman is assigned to be the helper? And the man is told that he can lie? The underpinnings are absurd, from a critical perspective, in that both the man and woman are given gender-coded physical and moral attributes.

The philosophical underpinnings of Turing's argument are unsound. One of the most compelling counterarguments was addressed by the philosopher John Searle in an argument known as the Chinese Room. Searle summarized it in a 1989 piece in the *New York Review of Books*:

A digital computer is a device which manipulates symbols, without any reference to their meaning or interpretation. Human beings, on the other hand, when they think, do something much more than that. A human mind has meaningful thoughts, feelings, and mental contents generally. Formal symbols by themselves can never be enough for mental contents, because the symbols, by definition, have no meaning (or interpretation, or semantics) except insofar as someone outside the system gives it to them.

You can see this point by imagining a monolingual English speaker who is locked in a room with a rule book for manipulating Chinese symbols according to computer rules. In principle he can pass the Turing test for understanding Chinese, because he can produce correct Chinese symbols in response to Chinese questions. But he does not understand a word of Chinese, because he does not know what any of the symbols mean. But if he does not understand Chinese solely by virtue of running the computer program for "understanding" Chinese, then neither does any other digital computer because no computer just by running the program has anything the man does not have.[3]

Searle's argument that symbolic manipulation is not equivalent to understanding can be seen in the popularity of voice interfaces in 2017. "Conversational" interfaces are popular, but they are far from intelligent.

Amazon's Alexa and other voice-response interfaces don't understand language. They simply launch computerized sequences in response to sonic sequences, which humans call verbal commands. "Alexa, play 'California Girls'" is a voice command that a computer can follow. *Alexa* is the trigger word that tells the computer that a command is coming. *Play* is a trigger word that means "retrieve an MP3 from memory and send the command *play* to a previously specified audio player, along with the MP3 file name." The interface is also programmed to capture whatever word comes after *play* and before the pause (the end of the command). That value is put into a variable such as *songname*, which is retrieved from memory and fed to the audio player. This process is procedural and unthreatening and shouldn't make anyone think that the machines are going to rise up and take over the world. Right now, a computer can't reliably distinguish whether it should respond to the previous command by playing Katy Perry's "California Gurls" or the Beach Boys' "California Girls." In fact, this exact problem is solved by running a popularity contest. Whichever song is played more often by all Alexa users is assumed to be the default choice. This is good for Katy Perry fans, but not so good for Beach Boys fans.

I'm going to ask you to keep the two competing ideas about narrow and general AI, and the idea of limitations, in your mind as you read. In this book, we'll stay squarely in the realm of reality: the world in which we have unintelligent computing machines that we *call* intelligent machines. However, we'll also look at how imagination—which is powerful and wonderful and exciting—sometimes confuses the way we talk about computers, data, and technology. I'd also ask you not to be disappointed like the art student at the science fair when you come up against what a colleague calls the *ghost-in-the-machine fallacy*—the reality that there is no little person or simulated brain inside the computer. There are different ways to react to this news: you can be sad that the thing you dreamed of is not possible—or you can be excited and embrace what *is* possible when artificial devices (computers) work in sync with truly intelligent beings (humans). I prefer the latter approach.

4 Hello, Data Journalism

We are at an exciting moment, when every field has taken a computational turn. We now have computational social science, computational biology, computational chemistry, or digital humanities; visual artists use languages like Processing to create multimedia art; 3-D printing allows sculptors to push further into physical possibilities with art. It's thrilling to consider the progress that has been made. However, as life has become more computational, people haven't changed. Just because we have open government data doesn't mean we don't have corruption. The tech-facilitated gig economy has exactly the same problems as labor markets have had since the beginning of the industrial age. Traditionally, journalists have investigated these types of social problems to create positive social change. In the computational world, the practice of investigative journalism has had to go high tech.

Many of the people who push the boundaries of what is technologically possible in journalism call themselves *data journalists*. Data journalism is a bit of a catchall term. Some people practice data journalism by making data visualizations. Amanda Cox, the editor of the *New York Times* section "The Upshot," is a master of this kind of visual journalism. Consider a 2012 story, "All of Inflation's Little Parts," which led to Cox winning the American Statistical Association's award for excellence in statistical reporting. The underlying data in the story is taken from the consumer price index, which is compiled by the Bureau of Labor Statistics monthly and used to measure inflation. In the graphic, a large half-circle is broken into colored mosaic tiles. The different sizes correspond to the percentages of Americans' spending.

A large shape, gasoline, represents 5.2 percent of spending. It is part of a category, transportation, that takes 18 percent of the average person's income. A smaller shape, representing eggs, is highlighted as part of the 15 percent of income spent on food and beverages. "High oil prices and drought in Australia are among the factors that have made food prices rise faster than they have since 1990," reads Cox's text. "Strong European demand for eggs has also affected prices in that category."[1] This text, and the intriguing shapes, open a window into the amazing ways that global citizens are connected via a complex web of trade. Eggs are global? Of course they are! Countries don't produce all of their own food anymore. Food is a global trade market. Western Australia contains a massive wheat belt. Australia overall exported $27.1 billion in food from 2010 to 2011 according to the Australian government's Department of Agriculture, Fisheries, and Forestry. The drought in the wheat belt led to less wheat production. Poultry feed in the United States is based primarily on cereal grains. Corn is preferred, but producers will use wheat if wheat is cheaper than corn. Less wheat available globally means that wheat prices are more expensive, which means that poultry feed producers either pay more for wheat or turn to also-expensive corn. As poultry farmers pay higher prices for chicken feed, they pass along the costs and charge higher prices for eggs. That increase is passed along to the consumer in the supermarket. The data provides a way to think through how a drought in Australia leads to higher egg prices at a North American supermarket, which is also a story about globalization, interconnectedness, and the environmental consequences of climate change. Cox uses her storytelling skills, her knowledge of how complex systems work in the world, her technological skills, and her keen design sense to create a visually exciting computational artifact that both informs and delights.

Other data journalists collect their own data and analyze it. In 2015, the *Atlanta Journal-Constitution* (AJC) gathered data on doctors who sexually abuse their patients. An AJC investigative reporter discovered that in Georgia, two-thirds of doctors who had been disciplined for sexual misconduct with their patients were permitted to practice again. This would have been enough for a story, but the reporter wondered if Georgia was typical or unusual. The story became a team investigation. The team gathered data from across the United States and analyzed more than one hundred thousand medical board orders from 1999 to 2015 related to disciplinary action against doctors. Their findings were shocking. All across the country, doctors were forgiven and allowed to resume practicing medicine after being found guilty of abusing patients. The very worst cases were horrifying. A pediatrician, Earl Bradley, was believed to have drugged over one thousand children with lollipops and molested them on video. He was indicted in 2010 on 471 charges of rape and molestation, and sentenced to fourteen life terms without parole. The AJC story led to awareness and positive reform, thankfully.[2]

In Florida, *Sun Sentinel* data journalists sat on the side of the highway and noted when police cars came by; they later requested the data from the police transponders at toll booths, and found that the police were systematically traveling at high speeds that endangered citizens. After the investigation, police speeding dropped 84 percent. This dramatic, positive public impact helped the story win the 2013 Pulitzer Prize for Public Service.[3] A lot of good data journalism comes out of Florida. For one thing, the narrative possibilities are endless. "Florida has long eclipsed California as the place where the bizarre, unusual and outlandish have become commonplace," Jeff Kunerth wrote in the *Orlando Sentinel* in 2013.[4] Everything that the US government does is public by default, but Florida has "sunshine laws" that guarantee that the public has access, and tapes, photographs, films, and sound recordings are also considered public records. Great open records laws mean that it's easy to get official government data, which means that a lot of data journalism is done in and about Florida.

Some data journalists get data from official sources and analyze it to find insights. These insights can lead to uncomfortable truths. For example, one example of a successful academic-industrial partnership came about when data journalist Cheryl Phillips of the Stanford Computational Journalism Lab organized a class project in which her students requested data on police stops from all fifty states. They analyzed the data for nationwide trends and released it online for reuse by other journalists. The Stanford journalists and all of the other journalists found that people of color were stopped far more often than white people in every state.[5]

Data journalism also includes algorithmic accountability reporting, which is the small corner of the field I occupy. Algorithms, or computational processes, are being used increasingly to make decisions on our behalf. Algorithms determine the price you see for a stapler when you're shopping online; they also determine how much you pay for health insurance. When you submit a job application or resume via an online job site, an algorithm generally determines whether you meet the criteria to be evaluated by a human or whether you're rejected outright. The role of the free press in a democracy has always been to hold decision-makers accountable. Algorithmic accountability reporting takes this responsibility and applies it to the computational world.

A prominent example of algorithmic accountability reporting is Pro-Publica's story "Machine Bias," published in 2016.[6] ProPublica reporters found that an algorithm used in judicial sentencing was biased against African Americans. Police administered a questionnaire to people who were arrested, and the answers were fed into a computer. The Correctional Offender Management Profiling for Alternative Sanctions (COMPAS) algorithm then spit out a score that "predicted" how likely the person was to commit a crime in the future. The score was given to judges in the hopes that the score would allow judges to make more "objective," data-driven decisions about sentencing. However, this resulted in African Americans receiving longer jail sentences than whites.

It's easy to see how technochauvinism blinded the COMPAS designers from seeing how their algorithm might be harming people. When you believe that a decision generated by a computer is better or fairer than a decision generated by a human, you stop questioning the validity of the inputs to the system. It's easy to forget the principle of garbage in, garbage out—especially if you really *want* the computer to be correct. It's important to question whether these algorithms, and the people who make them, are making the world better or worse.

The practice of using data in journalism is older than most people think. The first data-driven investigative story appeared in 1967, when Philip Meyer used social science methods and a mainframe computer to analyze data on race riots in Detroit for the *Detroit Free Press*. "One theory, popular with editorial writers, was that the rioters were the most frustrated and helpless cases at the bottom of the economic ladder, who rioted because they had no other means of advancement or expression," Meyer wrote. "The theory was not supported by the data."[7] Meyer conducted a large survey and performed a statistical analysis of the results using a mainframe. He discovered that the participants in the riots came from a variety of social classes. He won a Pulitzer for his reporting. Meyer called the application of social science to journalism *precision reporting*.

Later, when desktop computers came to every newsroom, reporters started using spreadsheets and databases to track data and find stories. Precision reporting evolved to what became called *computer-assisted reporting*.

Computer-assisted reporting is the type of investigative journalism used in the movie *Spotlight*, which dramatizes the *Boston Globe*'s Pulitzer-winning investigation into Catholic priests who were sexually abusing children—and the forces that were covering up the problem. To keep track of the hundreds of cases and hundreds of priests and parishes, the reporters used spreadsheets and data analysis. In 2002, this was state-of-the-art investigative practice.

As the Internet grew and new digital tools emerged, computer-assisted reporting evolved into what we now call *data journalism*, which encompasses visual journalism, computational journalism, mapping, data analysis, bot-making, and algorithmic accountability reporting (among other things). Data journalists are journalists first. We use data as a source, and we use a variety of digital tools and platforms to tell stories. Sometimes those stories are about breaking news; sometimes the stories are entertaining; sometimes the stories are investigative. They are always informative.

ProPublica, established in 2008, and the *Guardian* have been leaders in the field.[8] ProPublica, started by *Wall Street Journal* veteran Paul Steiger with philanthropic backing, quickly made a name for itself as an investigative powerhouse. Steiger has a deep investigative background: he served as the managing editor of the *Wall Street Journal* from 1991 to 2007, during which time members of the publication's newsroom staff were awarded sixteen Pulitzer Prizes. ProPublica reporters received the first of their many Pulitzer Prizes in May 2010. The organization's 2011 Pulitzer Prize for national reporting was the first such prize ever for stories not published in print.

Many Pulitzer projects have had data journalists, or people who identify as data journalists, on the team. Journalist and programmer Adrian Holovaty, who created the Django programming framework that's used by many newsrooms, published an online rant titled "A Fundamental Way Newspaper Sites Need to Change" in September 2006.[9] Holovaty advocated for newsrooms going beyond the traditional story model by integrating structured data into reporters' ordinary methods. His rant led Bill Adair, Matt Waite, and their team to create the PolitiFact fact-checking site, which won a Pulitzer in 2009. Waite wrote of the launch: "The site is a simple, old newspaper concept that's been fundamentally redesigned for the web. We've taken the political 'truth squad' story, where a reporter takes a campaign commercial or a stump speech, fact checks it and writes a story. We've taken that concept, blown it apart into its fundamental pieces, and reassembled it into a data-driven website covering the 2008 presidential election."[10]

Holovaty went on to make EveryBlock, a pioneering news app that integrated crime data and geolocation. It was the first to use the Google Maps API, leading Google to make the feature available to everyone.[11]

The *Guardian* broke ground in data journalism in 2009 when a team of reporters and programmers sought to review, via crowdsourcing, 450,000 records of expenses generated by members of Parliament. This effort was a follow-up to a scandal in which it was discovered that MPs were using government funds to pay for household and office expenses. The *Guardian* team also gained expertise in using computational methods to analyze large troves of leaked documents, as in their analysis of the Afghanistan and Iraq war logs.[12]

One important project in the field is an investigation by the *Wall Street Journal* into price discrimination.[13] Major chains like Staples and Home Depot were charging different prices on their websites depending on the zip code in which visitors seemed to be. The journalists used computational analysis tools to discover that customers in wealthier zip codes were being charged less than customers in poorer zip codes.

Academic research is an important complement to data journalism. Data journalists tend to rely on established scholarly research methods. Part of being a good journalist is knowing when to turn to a subject matter expert; another part is telling the difference between an expert and a shill. Data journalists synthesize expertise from a wide variety of fields. Georgia Tech professor Irfan Essa organized the first Computation + Journalism Symposium in 2008. At this annual event, journalists come together with researchers from communication, computer science, data science, statistics, human-computer interaction, visual design, and more to share their research and promote understanding. Northwestern professor Nicholas Diakopoulos, one of the conference co-founders, has written important works about reverse-engineering algorithms as part of holding decision-makers accountable. His paper "Algorithmic Accountability: Journalistic Investigation of Computational Power Structures"[14] describes some of his work and the work of other journalists in investigating algorithmic black boxes.

Though it is closely linked to computer science, data journalism is generally considered a social science. The most in-depth explorations of the field can be found in social science literature. C. W. Anderson published

"Towards a Sociology of Computational and Algorithmic Journalism" in 2012,[15] in which he united Schudson's four approaches to studying news with ethnographic insights gained from fieldwork at a Philadelphia newspaper between 2007 and 2011. Nikki Usher contributed additional ethnographic context with her book *Interactive Journalism: Hackers, Data, and Code*,[16] which is based on both fieldwork and interviews with data journalists at the *New York Times*, the *Guardian*, ProPublica, WNYC (New York Public Radio), AP, National Public Radio (NPR), and Al Jazeera English. Cindy Royal's work on journalists producing code[17] was important for understanding how journalists used code inside the newsroom, and it also prompted understanding of how journalism schools could integrate computational skills into their curricula. James T. Hamilton's 2016 book *Democracy's Detectives* outlined how crucial data-driven investigative journalism is for the public good—and how much this public service can cost. High-impact investigative data journalism stories cost hundreds of thousands of dollars to produce. "Stories can cost thousands of dollars to produce but deliver millions in benefits spread across a community," Hamilton writes.[18]

In 2010, Tim Berners-Lee gave the new field the computational stamp of approval when he said: "Journalists need to be data-savvy. It used to be that you would get stories by chatting to people in bars, and it still might be that you'll do it that way some times. But now it's also going to be about poring over data and equipping yourself with the tools to analyse it and picking out what's interesting. And keeping it in perspective, helping people out by really seeing where it all fits together, and what's going on in the country."[19] By the time Nate Silver launched FiveThirtyEight.com and published his book *The Signal and the Noise* in 2012, the term *data journalism* was in widespread use among investigative journalists.[20]

As computers have evolved, human nature has not. People need to be kept honest. I hope that this book will help you think like a data journalist so that you can challenge false claims about technology and uncover injustice and inequality embedded in today's computational systems. Using a journalist's skepticism about what can go wrong can help us move from blind technological optimism to a more reasonable, balanced perspective on how to live lives that are enhanced but not threatened or compromised by technology.

II When Computers Don't Work

5 Why Poor Schools Can't Win at Standardized Tests

Machines, code, and data can all work together to produce amazing, exciting insights. Getting hold of the right numbers can increase revenue, improve decision making, or help you find a mate—or so the thinking goes. The gospel of data is particularly fervent in the education world. In 2009, US Education Secretary Arne Duncan told a crowd of education researchers: "I am a deep believer in the power of data to drive our decisions. Data gives us the roadmap to reform. It tells us where we are, where we need to go, and who is most at risk."[1]

However, it is naive to believe that data alone can solve social problems. I learned this when I tried to use big data to help repair my local public schools. I failed—and the reasons why I failed have everything to do with why the current technocratic American system of standardized testing will never succeed.

One day, when my son was in first grade, I had trouble helping him with his homework. "I need to write down natural resources," he told me.

"Air, water, oil, gas, coal," I replied.

"I already put down air and water," he said. "Oil and gas and coal aren't natural resources."

"Of course they are," I said. "They're nonrenewable natural resources, but they're still natural resources."

"But they weren't on the list the teacher gave in class."

Parenting offers many moments when I feel out of my depth, but this moment felt like an epistemological dilemma. My own general knowledge (and the Internet) told me there were many possible "correct" answers. However, only one of these answers would get him full credit on the assignment.

I looked at the worksheet. It had a picture of a cow and an umbrella. "A cow is not a natural resource," I said.

"Animals are a natural resource," said my son.

"Cows are part of nature. They are not a natural resource."

"But the teacher *said* animals."

"Did the teacher also say umbrellas?"

"I think that means water. I already wrote that down."

"Let's look in the book," I said, getting frustrated. "If there's a worksheet, there's a book that goes with it."

"There's no book," said my son.

"Of course there's a book," I said.

"We're not allowed to take it home. It's only for in class."

"The teacher said at back-to-school night that there would be an online version of the book. Did she give you the website or the password?"

"No."

I spent the next hour trying to hack into the online textbook site or find a pirated copy online. No luck.

I knew my son would start taking standardized tests in third grade. If the first-grade homework was this confusing, I was really worried about how he—or any kid—was supposed to figure out the tests.

I had been spending time with civic hackers, the kind of people who build software and crunch government data for fun, and I decided to see if I could come up with a beat-the-test strategy derived from a popular SAT prep course I used to teach. In essence, I tried to game the third-grade Pennsylvania System of School Assessment (PSSA), the standardized test for the state in which we then lived. Along with a team of professional developers, I designed artificial intelligence software to crunch the available data.

In recent years, artificial intelligence has become increasingly valuable to journalists: Automated writing has helped journalists to more efficiently cover routine stories in sports and business. Machine learning has helped journalists to understand large datasets, resulting in document analysis tools like Overview Project or DocumentCloud. I was interested in a third dimension of artificial intelligence that I thought could help journalists to find stories in data: expert systems. As originally conceived, an expert system would act like a human expert in a box, dispensing advice. Unfortunately, this never worked. Cognition and expertise are too complex to be mapped onto automated procedures enacted by binary counting machines (which is ultimately what our present computers are). My research suggested, however, that this concept of an expert system could be successfully modified for public affairs reporting to enable journalists to discover stories in large public datasets quickly and efficiently.

I designed and built software to perform the necessary data analysis. I talked to teachers. I talked to students. I visited schools and sat through School Reform Commission meetings. After six months of this, I discovered that the test could be gamed. Not by using a beat-the-test strategy, but by a shockingly low-tech strategy: reading the textbook that contains the answers.

Philadelphia is the eighth-largest school district in the country, and its public students are overwhelmingly poor: in 2013, 79 percent of them were eligible for free or reduced-price lunch. The high-school graduation rate was only 64 percent and fewer than half of students managed to score proficient or above on the 2013 PSSA.

When a problem exists in Philadelphia schools, it generally exists in other large urban schools across the nation. One of those problems—shared by districts in New York, Washington, D.C., Chicago, Los Angeles, and other major cities—is that many schools don't have enough money to buy books. The School District of Philadelphia once tweeted a cheerful photo of former Philadelphia mayor Michael Nutter handing out two hundred thousand donated books to K–3 students. Unfortunately, introducing children to classic works of literature won't raise their abysmal test scores.

This is because standardized tests are not based on general knowledge. As I learned during my investigation, they are based on specific knowledge contained in specific sets of books: the textbooks created by the test makers.

All of this this is tied up with the economics of testing. Across the nation, standardized tests come from one of three companies: CTB/McGraw-Hill, Houghton Mifflin Harcourt (HMH), or Pearson. These corporations write the tests, grade the tests, and publish the books that students use to prepare for the tests. HMH had a 38 percent market share, according to its press materials. In 2013, the company brought in $1.38 billion in revenue.

That same year, Pennsylvania had a multi-million-dollar contract with a company called Data Recognition Corporation (DRC) to grade the PSSAs. DRC worked with McGraw-Hill as part of a consortium that had a $186 million federal contract to write and grade standardized tests for the rest of the country. McGraw-Hill, meanwhile, also wrote the books and curricula schools buy to prepare students for the tests. *Everyday Math*, the branded curriculum used by most Philadelphia public schools in grades K–5, is published by McGraw Hill.

Put simply, any teacher who wants his or her students to pass the tests must use books from the big three textbook publishers. If you look at a textbook from one of these companies and look at the standardized tests written by the same company, even a third grader can see that many of the questions on the test are similar to the questions in the book. In fact, Pearson came under fire in 2012 for using a passage on a standardized test that was taken verbatim from a Pearson textbook.

The issue often has as much to do with wording as it does with facts or figures. Consider this question from the 2009 PSSA, which asked third-grade students to write down an even number with three digits and then explain how they arrived at their answers. Figure 5.1 shows an example of a correct answer, taken from a testing supplement put out by the Pennsylvania Department of Education. Then, figure 5.2 shows an example of a partially correct answer that earned the student just one point instead of two.

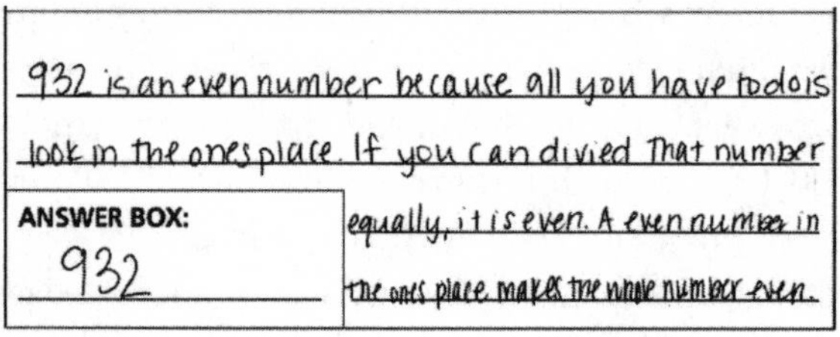

Figure 5.1
Correct answer on a question from the 2009 PSSA.

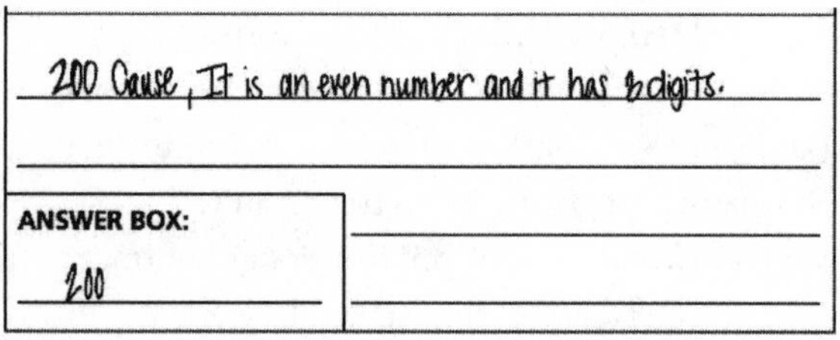

Figure 5.2
Partially correct answer on a question from the 2009 PSSA.

The second answer is correct, but the third-grade student lacked the specific conceptual underpinnings to explain why it was correct. The *Everyday Math* curriculum covers this rationale in detail, and the third-grade study guide instructs teachers to drill students on it: "What is one of the rules for odd and even factors and their products? How do you know that this rule is true?" A third grader without a textbook can learn the difference between even and odd numbers, but she will find it hard to guess how the test-maker wants to see that difference explained. In effect, these tests are for "narrow" intelligence, not general. This is a system that treats children like machine learners; if they are to spit out the "correct" answer when prompted (a questionable goal), they need to be inputted with the correct data: the data in the books.

Unlike college professors, who simply assign books and leave it to the students to buy them, K–12 teachers must provide students with books—but it's not a simple matter of ordering one book per student per subject. Based on the schools I visited and the teachers I interviewed, each student needs at least one textbook and one workbook per class, plus a bunch of worksheets and projects the teacher pulls from assorted websites (not to mention binder clips and construction paper and scissors and other materials for project-based learning). Books can be reused year to year, but only if the state standards haven't changed—which they have every year for at least the past decade.

Once I realized the direct connection between textbooks and standardized-test success, I tried to find out exactly how many Philadelphia schools were missing books from the big three publishers. I was also curious how much money it would take to make up for the shortfall.

The first challenge came when I asked the School District of Philadelphia for a list of which curricula were being used at which schools. If you want to know which books should be in a school, you need to know the name of the curriculum the school uses. (Using a branded curriculum like *Everyday Math* allows a school to place its orders more efficiently and negotiate a bulk discount.)

"We don't have that list," an administrator at the Philadelphia Office of Curriculum and Development told me. "It doesn't exist."

"How do you know what curriculum each school is using?" I asked.

"We don't."

There was silence on the phone for a moment.

"How do you know if the schools have all the books they need?"

"We don't."

According to district policy, every school is supposed to record its book inventory in a centralized database called the Textbook Storage System. "If you give me that list of books in the Textbook Storage System, I can reverse-engineer it and make you a list of which curriculum each school uses," I told the curriculum officer.

"Really?" she said. "That would be great. I didn't know you could do that!"

So, I did what computer programmers do in this kind of situation: I created a workaround. I built a program to look at each Philadelphia public school and see whether the number of books at the school was equal to the number of students. The results of the analysis did not look good. The average school had only 27 percent of the books in the district's recommended curriculum. At least ten schools had no books at all, according to their own records. Others had books that were hopelessly out of date.

I visited some of these schools and asked students how much access they had to textbooks. "We had books at my high school, but they were from, like, the 1980s," said David, a recent graduate of Philadelphia public schools. A junior at a public high school complained to me that her history textbook had pictures of testicles drawn on each page.

When I visited an algebra class at the Academy at Palumbo, a magnet school in South Philadelphia, a math teacher, Brian Cohen, seemed surprised by the information I presented to him. Palumbo's records showed that the school used a textbook called *Fast Track to a 5: Preparing for the AP Calculus AB and Calculus BC Examinations*. However, the quantity of books in the system read *0*.

"That's strange," said Cohen after I sat in on his Algebra I class. "I'm not sure why it says we have zero copies." Had that branded curriculum had been selected but never ordered? Or had the books had been ordered but intercepted somewhere along the way?

I asked if we could go look in the book closet, and Cohen took me down the hall. On the way, we stopped to chat with a colleague of his who taught calculus. "Do you have enough books?" Cohen asked.

"I do now," she said. "Some school in West Philadelphia closed, and I managed to get all the textbooks from there. I had a friend who hooked me up." However, she wasn't using *Fast Track to a 5*; she had a different calculus book that wasn't on my record sheet.

Urban teachers have a kind of underground economy, Cohen explained. Some teachers hustle and negotiate to get books and paper and desks for their students. They spend their spare time running campaigns on fundraising sites like DonorsChoose.org, and they keep an eye out for any materials they can nab from other schools. Philadelphia teachers spend an average of $300 to $1,000 of their own money each year to supplement their $100 annual budget for classroom supplies, according to a Philadelphia Federation of Teachers survey.

Cohen and I arrived at the math department's so-called book closet, actually just a corner inside the locked and empty office of the math department chairperson. "Here's where we keep the extra books," he said, gesturing to two short wooden bookshelves. A medium-sized box with open flaps sat on the floor. Cohen looked inside. "Well, we found the AP [Advanced Placement] Calculus books," he said. The box was filled with brand-new copies of *Fast Track to a 5*.

It would have been easy to blame this glitch on the lack of a centralized computer system. The only problem was that such a computer system did exist, and I was looking at a printout from it. The printout said Palumbo had zero copies of the book, but twenty-four books were sitting in front of me in a box on the floor of a locked office.

The Philadelphia schools don't just have a textbook problem; they have a data problem—which is actually a people problem. We tend to think of data as immutable truth, but we forget that data and data-collection systems are created by people. Flesh-and-blood humans need to count the books in a school and enter the numbers into a database. Usually, these humans are administrative assistants or teacher's aides. However, severe state funding cuts over the past several years have led to cutbacks in the school district's administrative staff. Even the best data-collection system is useless if there are no people available to manage it.

Michael Masch, the vice president of finance and chief financial officer at Manhattan College and the former chief financial officer of the School District of Philadelphia, told me that he used to routinely send his staff into schools to perform bookkeeping and other tasks that overworked principals couldn't handle. "Principals weren't good at managing cash accounts or student accounts. They needed support in performing administrative functions because they were understaffed," said Masch. "If the principal doesn't meet with every parent, deal with every crisis, they get criticized. If they don't do the invisible stuff, like the paperwork, they're not going to read about it in the newspaper. So, they triage."

When it comes to the book scarcity, Philadelphia principals react in predictable ways. "They are very possessive of their textbooks," Rebecca Dhondt, the parent of a second grader and a fourth grader at Jenks Elementary School, told me. "My daughter is not allowed to bring her textbook home because they don't want it to get lost." For the past two years, she has surveyed teachers to find out what's on their wish lists (mostly trade books and basic school supplies) and then collected donations from the community. "When I first did it last year, the principal said, 'Oh, we have some of that stuff,'" said Dhondt. As with the AP calculus books at Palumbo, the missing items were sitting somewhere in the school but hadn't made it into the right hands. "There's not enough support to connect the supplies in the supply closets or the libraries with the teachers in the classroom," Dhondt said. "They need to have enough money to connect the dots."

Keeping track of supplies is one problem; keeping track of the students who will use them is a whole other challenge. In Philadelphia schools, many students are in foster care or navigating other precarious living situations, which means they frequently switch schools. A report by the Children's Hospital of Philadelphia showed that one in five Philadelphia public high school students has been involved with the child welfare or juvenile justice system. One teacher told me that when she taught in a West Philly high school, she gained or lost a student at least every two weeks.

"There is a set of logistical issues in a district this big that most districts in the US don't face," explained Donna Cooper, the executive director of the Public Citizens for Children and Youth organization. "Everything isn't what it appears."

After I finished the first round of data analysis in 2013, I went to the school district and asked to present my findings to Philadelphia Superintendent William Hite. The district spokesperson told me Hite wasn't available and instead offered me a meeting with Stephen Spence, the deputy for the office of school-support services. Spence, a former gym teacher in his early sixties, was in charge of school openings and closings. His job used to be handled by an entire staff, but ever since the cutbacks, Spence had been taking care of everything from desks to carpets singlehandedly.

I asked him how he verified that schools had enough books at the beginning of each year. He explained that every principal was supposed to submit a school-opening checklist and a school-closing checklist. On that checklist (a Microsoft Word document that he emailed to all the principals), there was a box the principal could tick to indicate that the school had all the books it needed to operate.

"Inventory is not micromanaged at a central-office level," said Spence. "A principal that has very good skills with technology might develop an inventory system that they keep online. Another principal who is not so good with technology might have just a person who counts the books, carries them from one location to another, puts them in the closet, and visually checks that they're there."

I wondered about this, since a district-wide electronic system had been created several years back. In 2009, a student stood up at meeting of Philadelphia's School Reform Commission and proclaimed, "I don't have a book." After that, Superintendent Arlene Ackerman had resolved to computerize the District's inventory. Chief Information Officer Melanie Harris had told me that the system had been developed using internal resources.

"You're saying that the online system is no longer in use?" I asked Spence.

The principals preferred to use their own systems, he said, and report their inventory to him. "I rely on the principals and, I'm going to say, real-time data. It gets tracked through the documents we talked about previously: the school-opening and the school-closing checklists."

As Spence receives the principals' checklists, he enters the information into an Excel spreadsheet on his computer.

"Does this Excel document get shared with anyone?" I asked.

"It gets shared with assistant superintendents," said Spence. "We have meetings. We put the Excel spreadsheet on a projector on a large screen during our school-opening meetings."

As a data-science professional, it was clear to me that Spence was in over his head. Millions of books, hundreds of thousands of desks: it's impossible to keep track of all these objects without technology and sufficient people to track them. It's just as difficult to figure out how to use the data correctly.

The result is that Philadelphia's numbers simply don't add up. Consider the eighth grade at Tilden Middle School in Southwest Philadelphia. According to district records, Tilden uses a reading curriculum called *Elements of Literature*, published by Houghton Mifflin. In the 2012–2013 school year, Tilden had 117 students in its eighth grade, but it only had forty-two of these eighth-grade reading textbooks, according to the (admittedly flawed) district inventory system. Tilden's eighth grade students largely failed the state standardized test; their average reading score was 29.4 percent, compared with 57.9 percent districtwide.

One problem is that no one is keeping track of what these students need and what they have. Another problem is that there's simply too little money in the education budget. The *Elements of Literature* textbook costs $114.75. However, for 2012–2013, Tilden (like every other middle school in Philadelphia) was only allocated $30.30 per student to buy books—and that amount, which was barely a quarter the price of one textbook, was supposed to cover every subject, not just one. My own calculations showed that the average Philadelphia school had only 27 percent of the books required to teach its curriculum for 2012–2013, and it would have cost $68 million to pay for all the books schools need. Because the school district doesn't collect comprehensive data on its textbook use, this calculation could be an overestimate—but more likely, it's a significant underestimate.

At the end of the 2012–2013 school year, the book budget was eliminated altogether. In June 2013, the state-run School Reform Commission—which replaced Philadelphia's school board in 2001—passed a "doomsday budget" that fell $300 million short of the district's operating costs for the 2014 fiscal year. (The governor of Pennsylvania already had cut almost a billion dollars from public education funding in 2011.) Philadelphia schools were allotted zero dollars per student for textbooks. The 2015 budget likewise featured no funding for books.

This kind of complex bureaucratic morass was perfect for understanding with technology. Modeling the bureaucratic standards in code, then using the data to measure how well the system was meeting its own standards—this is pretty much what computers are built for. This kind of technique could be used for any other complex bureaucratic process as well. However, it also highlights the complexity of our social and public systems.

During this project, I gained a lot of appreciation for the Bill & Melinda Gates Foundation, the education funding powerhouse that was a prime mover on the development and adoption of Common Core State Standards. In a sense, the situation at all public schools is an engineering problem. The state standards are like blueprints for a house. If you have blueprints, and all the stakeholders have agreed on the blueprints, a contractor can go ahead and build the house specified in the blueprints. However, if the blueprints change, the contractor has to shift the deadline. Every time a new feature is added to the house, the cost increases and the deadline moves further away.

The same phenomenon happens in software projects. Usually, the people in charge neglect to specify all the features they need in a software system at the beginning of the project. Every time a new software feature is added, it increases the cost and pushes the launch deadline out. It's called *scope creep*. As the head of Microsoft, Bill Gates probably led more large-scale software development efforts than anyone else. Viewed this way, it makes perfect sense that the Gates Foundation would have stepped in to finalize the standards, get states to agree to the standards, and then implement curricula and testing to measure performance relative to the standards. It's an elegantly engineered solution. If people had managed to agree on the standards and the standards didn't change for a period of years, it might have worked.

However, this was not just an engineering problem: it was a social issue. Educational standards are not natural laws. They are ideas that arise from specific political and ideological contexts. For example, elected officials on school boards in Texas and California likely have different perspectives on how prominently evolution and climate change should figure in the standard curriculum. California school boards likely disagree with each other, as do Texas districts. If you want to see the kinds of fireworks that can explode around educational standards, check out the 2014 Colorado fights about the content of the AP US History exam.[2] Republicans, Democrats, the National Education Association, the American Federation of Teachers, Americans for Prosperity (a conservative group backed by the libertarian Koch brothers), school choice activists, real estate developers, elected officials of local school boards, and the College Board all entered into in this skirmish that made national news. Education has long been a battleground in America's culture wars.

The case of the school standards failure nationwide is an illustration of what happens when engineering solutions are applied to social problems. These formal engineering solutions can become so complex and time-consuming and data-driven that the quest for data or better technology obscures the social issues that are at stake.

Engineering solutions are ultimately mathematical solutions. Math works beautifully on well-defined problems in well-defined situations with well-defined parameters. School is the opposite of well-defined. School is one of the most gorgeously complex systems humankind has built. I go into my classroom every day and leave surprised by what has transpired. My students have endlessly complex lives. They have other deadlines, family drama, travel plans. Occasionally, they must meet the demands of their own children. It's an unpredictable environment. This is part of what I love about teaching; I'm a bit player in helping these students to develop into the best people they can be.

But, sometimes, the unpredictability also irks me as a computer scientist. When I make a syllabus for a class, I'm making the blueprint for the semester. If everyone followed the rules and the schedule in the syllabus, the semester would proceed in an orderly fashion and everyone would learn—but not once in a decade of teaching have I progressed through the assignments and readings in the syllabus without making adjustments.

To be clear: these adjustments make the class better. I adjust my syllabus to accommodate and enhance the experience of the students in the room. If most students don't seem to have mastered a foundational computational or journalistic concept, I back up and teach it to them before we move on to more advanced topics. This is consistent with current pedagogical best practices and evidence-based educational research. Tailoring the curriculum to the students in the room makes the class experience better. However, the engineer inside me sighs every time I adjust the orderly grid that maps out exactly what the students are to read and what assignments are due.

I can easily adjust my syllabus because I teach roughly thirty students each semester, and all the students are highly motivated. The students come to class once or twice a week, and they don't tend to take my classes—which are both quantitative and qualitative—unless they really want to. If I were teaching at a public K–12 school, however, the situation would be different. K–12 teachers also personalize the syllabus—but they're often managing more than thirty students in the classroom all day, five days a week, and the students *have* to be there instead of *wanting* to be there. Because I teach at a private university, I can ask my students to buy a new book in the middle of the semester if it means they will have an enhanced learning opportunity. At a public K–12 school, the school provides the books. Few teachers are allowed to buy thirty-plus copies of a book in the middle of the semester, and even if they were, the procurement process would take weeks. Implementing change in my small college classroom is like turning a high-performance sports car. Implementing change in a public K–12 class-room is like turning a cruise ship that's running at high speed.

A technochauvinist might look at this problem and suggest solving it with technology. We could make all the books available electronically and have the students access the books on their phones, because all students have phones.

Wrong.

Phones are great for reading short works, but long works are difficult and uncomfortable to read on a phone. Studies show that reading on a screen is worse than reading on paper in an educational context. Speed, accuracy, and deep learning all suffer when research subjects read on screens. Paper is simply a superior technology for the kind of deep learning that we want students to engage in as part of their education. Reading on a screen is fun and convenient, yes—but reading for comprehension is not about fun or convenience. It's about learning. When it comes to learning, students gen-erally prefer paper to screens.[3]

Another technochauvinist might suggest giving all the students iPads or Chromebooks or some kind of e-reader and making all the books available electronically: another good idea, but the obstacles are obvious. Have you ever been around a child or teenager? Have you noticed how often they lose things? Kids lose gloves, hats, keys, tablets—everything, really. Kids break things constantly. So, in a school of five hundred or a thousand or more kids, it doesn't make sense to buy everyone a $200 tablet or computer, then need to replace it at the same rate. Computers in a lab or in the hands of heavy users have a lifespan of about two years, if that.

A book has a lifespan of five-plus years. A paperback book for a language arts class costs as little as $0.99. The book doesn't need maintenance and doesn't need upgrading. It's cheap to replace. A computer, on the other hand, comes with infrastructure needs. If you get computers for five hun-dred students, you're not just buying computers. You're also committing to a constellation of associated services, costs, and maintenance.

Let's say you are an administrator for a school that wants to implement a one-to-one laptop program for five hundred students. You also need to provide twenty-four-hour telephone and email support for all five hundred students and fifty-plus teachers and twenty-plus administrators and staff, plus all the parents of all the students. You must adapt the electrical sys-tems in the aging school building to make sure you have enough power to plug in the computers. You must upgrade the school HVAC to make sure there's air conditioning, because computers generate a lot of heat and will malfunction in too-hot conditions. You need a powerful, always-up Wi-Fi network that can handle the bandwidth needs of over six hundred people streaming data from 6:00 a.m. to 7:00 p.m. daily. That Wi-Fi network has to work everywhere, even in classrooms that were built in the 1950s using solid cinderblock walls that deaden Wi-Fi signals. You must administer the passwords for the Wi-Fi network. You must help people who forget their passwords—even at 3:00 a.m. You need to add a secure network infrastructure for a learning-management system that teachers can use to distribute assignments or communicate with students and parents. You need that secure network infrastructure to comply with the Family Educational Rights and Privacy Act (FERPA), which protects the privacy of educational records but is ambiguous about what constitutes *educational records*. You need to negotiate agreements and workarounds to accommodate the Children's Online Privacy Protection Act (COPPA), which says that companies must obtain parental consent and protect the personal information of children younger than thirteen years, even though children start school at six or seven and are using computers in school before age thirteen. You need to set up IDs for hundreds of students and teachers and staff. You must shut down the IDs for all the students and staff who leave. You must check out the electronic books at the beginning of the semester and check them back in at the end of the semester. You need to get licenses for the electronic books, pay for the licenses, keep the licenses up-to-date, and deal with the inevitable problems of some students not having the right licenses and not being able to do their homework. You must deal with firewalls. You must deal with the situation of students who don't have Internet access at home: How are they supposed to do their homework? You need to repair the computers when they break, every day, and have a stockpile of loaners for computers that need more extensive repairs. You need extra chargers for the students who accidentally leave their charging cords at home. You need to replace lost computers. You need to replace stolen computers. You must develop acceptable use policies that discourage students from viewing inappropriate content, whereas what you really mean is "don't use the school-provided computers to look at porn, ultraviolent videos, drug content, or anything else that would make your parents upset with us." You must make the students (and teachers and staff) comply with these policies and create a culture in which people voluntarily comply rather than being forced to do so. You must get this entire infrastructure up and running, then maintain it 24-7 for the indefinite future, plus adopt new technologies as they arise. You must do this with minimal trained staff, minimal budget, and a tiny salary.

The challenges don't stop there. As teachers and administrators learned from the $20 million One Laptop Per Child (OLPC) initiative announced in 2005, simply handing students laptops doesn't mean they will use them for education. An OLPC deployment in Paraguay found that unless there were specific instructions given for how the computers should be tied into curricular activities, both teachers and students merely used the laptops for recreational purposes like playing games or watching videos.[4] Teachers needed training and support in order to integrate computers into their lesson plans—and even then, the logistical problems (breakage, loss, etc.) proved to be insurmountable obstacles. A minor scandal erupted when adults discovered that kids at an OPLC school in Nigeria were using the laptops to look at porn.[5]

It's not easy. Given all of this, it seems cheaper—and easier—to just use books.

I first wrote about this problem in 2014.[6] After the story ran, things changed—very slowly. The budget for the 2017 school year finally featured a proposal for new books. "Proposed investments for FY16 and FY17 include $32 million of instructional material refresh and $12 million for counselors and nurses," read a line in the 2017 School District of Philadelphia budget. Philanthropic organizations donated $10 million to "install leveled libraries in all K–3 classrooms." This was part of a $440 million investment in K–8 reading and math materials, high school technology updates, more AP and gifted opportunities, and counselors and nurses in every school.[7]

If we look at this through the lens of Internet speed, which is supposed to be lightning-fast, this tech project took far too long. It took six months to develop and two years for the change to be seen in the world. A techno-chauvinist would say that waiting two years for social change is too long. However, changing a large bureaucracy is hard. Like turning a cruise ship at full power, it takes a while; there's no way around it. The same is true for other algorithmic accountability investigative reporting. It takes a long time to do, you don't necessarily know what you're going to find, and the social impact may not happen for years.

I'm trying to be optimistic about the "instructional material refresh" planned for Philadelphia. I'm not convinced that it will be effective for more than a year—in part because the standards are likely to change again, requiring different materials; the test vendor is likely to change again, requiring teachers to shift their teaching strategies; and so on. I hope that the standards stop changing for at least a few years so that teachers, students, and school systems can catch up and implement the best practices that work for their specific school populations. I hope that state governments decide to fully fund public school systems, from pencils to laptops in the classrooms, and up to and including tater tots in the cafeterias. I want to see public education succeed. I want our next generation to be tech-savvy, empowered young people who learn civics and art and literature and math and computer science and statistics and history and all the marvels of the world around us. I want them to be full participants in the American experiment and the American dream. I'm not convinced, unfortunately, that I'll get what I want in this case.

6 People Problems

It may seem like ideas about education and digital technology stem from many different authors and thinkers, but digging deep reveals that most such ideas come from a single small group of elites who have been imagining and misunderstanding the interplay between technology and social issues since the 1950s. Understanding the deep connections these people have to each other can help us push back against too-simple, dysfunctional thinking about technology.

Computer systems are proxies for the people who made them. Because there has historically been very little diversity among the people who make computer systems, there are beliefs embedded in the design and concept of technological systems that we would be better off rethinking and revising. To see the consequences of this insular thinking, consider a story about tech gone wrong:

It was a clear day in late July 2016, and David Boggs thought it was perfect weather for flying. Boggs had just gotten a new toy: a drone, equipped with the latest in streaming-video technology. He was eager to test it out, so when his friends came over, he pulled out the drone and showed them what it could do.

The drone flew up, down, and around the yard. Boggs and his friends cheered it on as they watched the flight footage on an iPad. Their small Kentucky town, in Bullitt County, just outside Louisville, looked different from the air. The neat one- and two-story houses were reduced to the size of dollhouses. As the drone flew higher, the peaked roofs of the town subdivisions turned to gray rectangles. The wooded area near Boggs's house looked like a green river sweeping through the neighborhood. The fields looked vast.

Boggs directed the drone west across Highway 61 and then turned north, intending to take footage of a buddy's house. There was a loud bang: the drone lost altitude rapidly and went still on the ground.

The Merideth kids had been playing outside when they heard a loud droning noise. Drones work like helicopters, but they sound different. A helicopter has a loud thumping noise similar to a bass drum. A drone gives off a high-pitched keening noise, as if a very small child is yelling "EEEEEEEEEE!!!!!!" at the top of her lungs without stopping. A NASA study found that the noise of a flying drone is far more annoying than noise from a vehicle on the ground.[1] The Merideth kids heard the noise and ran to and told their dad, Willie. Everyone was confused. Was it a predator drone? Were the kids in danger? The noise continued, making it hard to think. Willie Merideth grabbed his shotgun, loaded it with birdshot, and fired at the flying target. The drone veered crazily and crashed out of sight in a nearby park. The noise stopped.

Boggs and his friends drove to the crash site shown on the iPad. They saw Merideth in his front yard, agitated. Everyone realized what had happened. Boggs was pissed. He had paid $2,500 for the drone, and his neighbor just *shot* it out of the sky? Merideth was also pissed. What was his neighbor doing, spying on his family from the air? Wasn't this a free country, where citizens have a right to privacy in their own homes? The situation escalated. Merideth pointed his shotgun at Boggs and his drone crew. Boggs called the police. The officers arrived, and they didn't know what to do—there was nothing on the books about how to mediate a dispute between citizens over a flying robot. It wasn't out-of-season hunting, because the drone wasn't an animal. It wasn't willful destruction of another person's property, because Merideth was on his own property when he shot down the drone.

Eventually, the officers decided to arrest Merideth because he was the one with the gun. At the station, they charged him with first-degree endangerment for shooting into the air and criminal mischief. His wife posted his $2,500 bail, and he was home shortly thereafter. A few months later, a judge dismissed the charges, ruling that Merideth was within his rights to shoot a robot that had hovered over his property inappropriately, invading his privacy.[2]

I have a different take on the situation. I want to ask the drone designers and marketers: *What did you think was going to happen?* America is a heavily armed country. You make a flying spy robot that emits an annoying noise, and you make virtually no rules or establish any guidance or social norms about using the robot or its video camera. Did you think about what possibly could go wrong?

This naïveté about the inevitable problems that arise when people use new gadgets shows up again and again in tech culture. Invariably, there are negative social consequences. Microsoft developers created a Twitter bot, Tay, that was intended to "learn" from its direct interactions with other Twitter users. Twitter users quickly demonstrated why Twitter has a reputation as a platform rife with abuse and harassment by directing a tidal wave of filth at Tay. The bot "learned" to spout white supremacist hate speech. Developers, surprised, shut it down.[3]

Another time, developers wanted to demonstrate the kindness of strangers by creating a GPS-enabled doll, hitchBOT, that was supposed to hitch rides all over the country. The idea was that you'd pick up hitchBOT, take it to your next destination, and leave it for someone else to pick up. In this way, hitchBOT would journey all over the country and have nice experiences with nice people who like to help others with technology. HitchBOT made it as far as Philadelphia, where the doll was dismembered and left in a dark alley.[4]

Blind optimism about technology and an abundant lack of caution about how new technologies will be used are the hallmarks of technochauvinism.

The story of how tech creators ended up with a reckless disregard for public safety and the public good starts with my favorite tech titan, Marvin Minsky. Minsky, a Harvard and Andover and Princeton grad, was an MIT professor and is usually considered the father of artificial intelligence. Look behind the scenes at the creation of virtually any high-profile tech project between 1945 and 2016, and you'll find Minsky (or his work) somewhere in the cast of characters.

Minsky's lab at Massachusetts Institute of Technology (MIT) is where hackers were born. It was terribly informal. Minsky's first recruits came from an MIT student group called the Tech Model Railroad Club (TMRC), whose members were building their own relay computers to power their model trains. The TMRC members were absolutely crazy about tinkering with machines. MIT had one of the few mainframe computers in the world at the time, in the late 1950s, and the TMRC members regularly snuck into the mainframe room after hours to play with it and run homemade programs.

Some professors might have disciplined students for breaking in and illicitly using university resources. Minsky hired them. "These were weird people," he recalled in an oral history.[5] "They had an annual contest to see who could ride every New York subway in the shortest time. It takes something like thirty-six hours. People would log these things very carefully and would study the schedules and plan their whole trip. These people were nuts." It was a productive kind of nuts for computer science, however. This obsessive attention to detail and insatiable desire to build things turned out to be exactly the right characteristics needed to write computer programs and build hardware. Minsky's lab flourished.

His recruiting method was unorthodox. Because Minsky was Minsky— the kind of person who always had a graduate student or a visitor living upstairs, who if you sat in his living room for long enough a politician or a science fiction writer or a famous physicist would drop by to chat—he never had to actively recruit. "Somebody would send a message or a letter to say: 'I'm interested in this,' and I'd say: 'Well, why don't you come here and see how you like working here?'" Minsky recalled. "The person would come for a week or two, we'd pay them enough to live on, and then they'd go away if they didn't hit it off. I don't remember ever making a decision to tell someone to go away. It's really very bizarre, but this was a self-energizing community. These hackers had their own language. They could get things done in three days that would take a month. If somebody appeared who had the talent, the magic touch, they would fit in." The TMRC and Minsky's lab were later immortalized in Stewart Brand's *The Media Lab* and Steven Levy's *Hackers: The Heroes of the Computer Revolution*, in addition to many other publications.[6] The hacker ethic is also what inspired Mark Zuckerberg's first Facebook motto: "Move fast and break things." Minsky was part of Zuckerberg's curriculum at Harvard.

Minsky and a collaborator, John McCarthy, organized the very first conference on artificial intelligence, at the Dartmouth Math Department in 1956. The two went on to found the Artificial Intelligence Lab at MIT, which evolved into the MIT Media Lab, which remains a global epicenter for creative uses of technology and has generated ideas for everyone from George Lucas to Steve Jobs to Alan Alda to Penn and Teller. (The MIT Media Lab was also kind enough to employ me for a software project devoted to Minsky's theories.)

Minsky's career was marked by good fortune at every turn. Most scientists today have to hustle for funding in an ever-shrinking grant environment. Minsky was of the generation that had money flowing out of the taps. He said in an oral history:

Until the 1980's, I never wrote a proposal. I just was always in the environment where there would be somebody like Jerry Wiesner of MIT.

John McCarthy and I had started working on artificial intelligence in about 1958, or 1959, when we both came to MIT. We had a couple of students working on it. Jerry Wiesner came by once and said, how are you doing? We said, we're doing fine, but it would be nice if we could support three or four more graduate students. He said, well, go over and see Henry Zimmerman and say I said that he should give you a lab. Two days later we had this little lab of three or four rooms, and a large pile of money that IBM had given to MIT for the advancement of computer science and nobody knew what to do with it. So they gave it to us.

A large pile of money, some endlessly creative mathematicians with speculative ideas about what might be possible in the future: this was how the field of artificial intelligence began. Eventually, Minsky's small, elite circle of individuals came to dominate the technological conversation in academia, industry, and even Hollywood.

When science fiction author Arthur C. Clarke worked with Stanley Kubrick to develop *2001: A Space Odyssey*, he turned to his friend Minsky for advice on how to imagine an artificial superintelligence on a spaceship that tries to save the world and ends up destroying the crew. Minsky delivered. Together, they created HAL 9000, a computer that even today embodies all of the promise and terror of what machines might do. Most people remember HAL's single, glowing red "eye." That ominous eye is almost identical to an eyeball (actually, a display unit) on ENIAC, which is considered the world's first programmable, general-purpose digital computer. John von Neumann, who came up with one of the core concepts of computer storage that led to ENIAC, was one of Minsky's mentors.

Minsky's literary taste ran almost exclusively to science fiction. He wrote his own, and he was also friends with Isaac Asimov and other prominent science fiction writers. Sometimes in the friendships, the lines between science fiction and reality became blurred. Minsky talked about some of their wacky projects in an interview:

One of the things I was interested in was Arthur Clarke's idea of a space elevator. I must have spent about six months working with some scientists at Livermore who were thinking about designing such things. It's possible, in principle, to build a kind of pulley—a belt—made of carbon fiber, some incredibly strong piece of wire and have this go up from earth to something higher than a synchronous satellite and down again. So you could have a pulley that would haul things into space, and Arthur Clarke had worked out the theory of that. He called it a fountain.

To recap: Clarke, a science fiction writer, imagined an elevator fountain that goes to outer space. The writer convinced his scientist friend Minsky (in whose house he lived occasionally) that the space elevator would be a good idea. Minsky convinced some of his friends at the Lawrence Livermore National Laboratory, a defense contracting site that is today funded by the National Nuclear Security Administration and the Department of Energy, to explore the idea of making a gigantic outer space dumbwaiter. And these eminent scientists worked on the dumbwaiter to outer space for an *entire six months.*

Everybody in tech knew Minsky, and everyone relied on him. Steve Jobs famously got the idea for a computer with a mouse and a GUI from Alan Kay and his team at Xerox PARC. When Jobs left Apple in 1985 and John Sculley took over, Kay told Sculley they needed to go out and find sources of new technology. They weren't going to be able to turn to PARC for Apple's next big move, Kay said. "That led to us spending a lot of time on the East Coast, at the Media Lab in MIT, where we worked with people like Marvin Minsky and Seymour Papert," Sculley said in a 2016 interview.[7] "A lot of that technology we ended up putting into a concept video Alan and I produced, called 'Knowledge Navigator.' That predicted that computers were going to become our personal assistants, which is what's happening right now," with voice assistant technology like Apple's Siri, Amazon's Alexa, and Microsoft's Cortana.

All these assistants are given female names and default identities by tech executives and developers—no accident. "I think that probably reflects what some men think about women—that they're not fully human beings," said social anthropologist Kathleen Richardson, the author of *An Anthropology of Robots and AI: Annihilation Anxiety and Machines* in a 2015 interview with LiveScience. "What's necessary about them can be replicated, but when it comes to more sophisticated robots, they have to be male."[8]

Minsky's world view is even behind the scenes in the founding of Internet search, which most of us use every day. As PhD students at Stanford, Larry Page and Sergei Brin invented PageRank, the revolutionary search algorithm that led to the two founding Google. Larry Page is the son of Carl

Victor Page Sr., an artificial intelligence professor at Michigan who would have read Minsky extensively and interacted with him at AI conferences. Larry Page's PhD advisor at Stanford was Terry Winograd, who counts Minsky as a professional mentor. Winograd's PhD advisor at MIT was Seymour Papert—Minsky's longtime collaborator and business partner. A number of Google executives, like Raymond Kurzweil, are Minsky's former graduate students.

Minsky was a Gladwellian connector. As far back as the 1950s, there were only a handful of places in the whole country of millions of people where computing machines were—and Marvin Minsky was in all of these places, hanging around, doing math, and building things and tinkering and hanging out.

Minsky-style creative chaos is fun and delightful and inspirational. It's also dangerous. Minsky and his generation did not have the same attitudes toward safety that we know are important today. There was a kind of casual disregard for radiation safety, for example. Once, a computer scientist and former Minsky graduate student named Danny Hillis showed up to Minsky's house with a radiation detector in his pocket. (Hillis, a supercomputer inventor, now runs the Long Now Foundation with *Whole Earth Catalog* founder Stewart Brand; the foundation is devoted to building a mechanical clock that will run for ten thousand years in a cave on a Texas ranch owned by Amazon founder Jeff Bezos.) The radiation detector started going crazy. Hillis, who had lived with the family for a time, poked around the house to find the source of the radiation. The alarm seemed loudest next to a closet. Hillis opened it and found the closet stuffed full of chemicals. He removed each one, but nothing was the source. Then, he found a secret panel in the back of the closet. Intrigued, he popped it open and found a human skeleton.

Hillis ran upstairs to tell Minsky and his wife Gloria Rudisch about the find. They were more excited than surprised. "Is that where that is?" Rudisch said. "We've been looking for that thing for years." It was a skeleton that she had used in medical school. It also wasn't the source of the radiation, however.

Eventually, Hillis excavated more stuff out of the closet and found a lens from an old spy camera that Minsky had gotten at a surplus store. Lenses of that vintage were sometimes treated with radioactive elements to increase the index of refraction. "It was dangerously radioactive," Hillis recalled.

"I got it out of the house."[9] When it came to tinkering, many makers of Minsky's generation felt that conventional rules didn't apply to them. For example, Minsky liked to tell a story about some friends of his who built an intercontinental ballistic missile (ICBM) in the backyard of a house that once belonged to architect Buckminster Fuller.

This attitude, that creating mattered more than convention (or laws), was what people of Minsky's generation passed on to their students. It shows up later in the behavior of tech CEOs like Travis Kalanick, who in 2017 was ousted from his top position at Uber for (among other things) creating a culture of sexual harassment. Kalanick also had the attitude that laws didn't matter. He launched Uber in cities worldwide in defiance of local taxi and limousine regulations, created a program called Greyball to help Uber computationally evade sting operations by law enforcement, was captured on camera verbally abusing an Uber driver, and looked the other way when Uber drivers raped passengers.[10] According to a blog post by former Uber engineer Susan Fowler, Kalanick's tech managers were almost cartoonishly incompetent at dealing with the harassment complaints Fowler lodged. Fowler was routinely passed over for promotion and was sexually propositioned by male coworkers. Uber's HR team should have recognized that Fowler was facing a textbook case of gender bias in the workplace. Instead, they put her on probation and told her it was her fault.

Disregard for social convention goes back farther than Minsky, back to computing pioneer Alan Turing, who, like Minsky, did his graduate work at Princeton. Turing was hopeless at social interaction. Turing's biographer—Jack Copeland, director of the Turing Archive for the History of Computing—writes that Turing preferred to work in isolation: "Reading his scientific papers, it is almost as though the rest of the world—the busy community of human minds working away on the same or related problems—simply did not exist."[11] Unlike the character portrayed by actor Benedict Cumberbatch in the Turing biopic *The Imitation Game*, the real Turing was slovenly in appearance. He wore shabby clothing, his fingernails were always dirty, and his hair stuck out at wild angles. Copeland writes:

Once you got to know him Turing was fun—cheerful, lively, stimulating, comic, brimming with boyish enthusiasm. His raucous crow-like laugh pealed out boisterously. But he was also a loner. "Turing was always by himself," said codebreaker Jerry Roberts: "He didn't seem to talk to people a lot, although with his own circle he was sociable enough." Like everyone else Turing craved affection and company, but he never seemed to quite fit in anywhere. He was bothered by his own social strangeness—although, like his hair, it was a force of nature he could do little about. Occasionally he could be very rude. If he thought that someone wasn't listening to him with sufficient attention he would simply walk away. Turing was the sort of man who, usually unintentionally, ruffled people's feathers—especially pompous people, people in authority, and scientific poseurs. He was moody too. His assistant at the National Physical Laboratory, Jim Wilkinson, recalled with amusement that there were days when it was best just to keep out of Turing's way. Beneath the cranky, craggy, irreverent exterior there was an unworldly innocence though, as well as sensitivity and modesty.

Notice the phrase "once you got to know him." That's what you say about someone who's unpleasant or unbearable, but there is some reason that you have to look past the person's awfulness. In Turing's case, most people pardoned his behavior because he was mathematically brilliant.

This looking beyond superficial features like physical appearance is one of the wonderful things about the social culture of mathematics. However, it's also a drawback when that same disdain for social conventions leads to a valorization of mathematical ability over social fabric. Disciplines like math, engineering, and computer science pardon a whole host of antisocial behaviors because the perpetrators are geniuses. This attitude forms the philosophical basis of technochauvinism, in which efficient code is prioritized above human interactions.

Tech also inherited mathematicians' worship of the cult of genius. This cult of genius has led to much mythologizing; it also enforces the boundaries of the industry and camouflages a range of structural discrimination. Math is obsessed with pedigrees. There is a popular online math genealogy project that's a crowdsourced list of mathematicians and their "ancestors" and "descendants," organized according to who got their PhD where and under whom. Minsky's intellectual "ancestry" can be traced in an unbroken line all the way back to Gottfried Leibniz in 1693. To understand why this matters, we need to walk through the development of the modern-day computer.

The earliest computing machine, as you probably recall from elementary school math class, was the abacus. The abacus is a base-ten counting device, because humans have ten fingers and ten toes. The abacus, commonly seen today as a set of ten beads strung on wires, was what people used for calculation for centuries.

The next major development in mathematical technology was the astrolabe, used for celestial navigation at sea. Then came a variety of clocks:
water-powered, spring-powered, and mechanical. Although these were all
important and ingenious inventions, the major innovation in terms of
computer design came in 1673 when Gottfried Leibniz, a German lawyer
and mathematician, built a device called a *step reckoner*. The step reckoner
had a set of revolving gears that moved via a crank. Once you passed nine
on a gear, that gear reset to zero and the adjacent gear incremented by one.
Each gear was a "step" representing an increment of ten. This design was
used to build calculating machines for the next 275 years.[12]

Leibniz had no time for mere arithmetic; he had more important
math to do. After he invented his machine, he famously said: "It is beneath
the dignity of excellent men to waste their time in calculation when
any peasant could do the work just as accurately with the aid of a
machine."

When Joseph Marie Jacquard released the punch-card loom in 1801, it
got mathematicians thinking differently about machines that might help
calculate. Jacquard's loom ran on binary logic: a hole in the card meant
binary one; no hole meant binary zero. The machine wove its intricate patterns based on whether there was a hole or not.

It took a few decades for people to figure out the details, but finally
there was a breakthrough in in 1822, when English scientist Charles Babbage began work on what he called a *difference engine*. This machine could
proximate polynomials, meaning it allowed mathematicians to describe
the relationship among several variables, such as range and air pressure.
The difference engine was also designed to compute logarithmic and
trigonometric functions, which are unpleasant to calculate by hand. Babbage worked on building the difference engine for years, eventually using
twenty-five thousand components that together weighed fifteen tons, but
he never got it to work. However, in 1837, Babbage published another, better idea: an *analytical engine*. This was a design for a machine that could
interpret a programming language with conditional branching and loops;
it had features recognizable from today's computers, like the ability to perform arithmetic and process logic and add memory. Ada Lovelace, generally
considered the first computer programmer, wrote programs for this hypothetical machine. Unfortunately, the analytical engine was so far ahead of
its time that it didn't work either. Scientists assembled it from Babbage's

designs in 1991 and discovered that it would have worked—if there were important other components, like electricity.

The next milestone toward the development of the modern computer happened when English mathematician and philosopher George Boole proposed Boolean algebra in 1854. Based on work by Leibniz, Boolean algebra is a logic-based system in which there are only two numbers, 0 and 1. Calculations are achieved via two operators: AND or OR.

As the nineteenth century progressed, mechanical adding machines became more sophisticated. William Seward Burroughs (grandfather of beat novelist William S. Burroughs) made a fortune from a mechanical adding machine he patented in 1888. After Thomas Edison released his first light bulb in 1878, electricity became widely available and revolutionized every kind of machinery. New electromechanical advances meant that anyone could do addition and subtraction and multiplication and division on an adding machine. However, it required pressing a lot of buttons repeatedly, and it was laborious. Human computers were still essential to the project of higher mathematics.

A *human computer* was a person, a kind of clerk, who was hired to perform calculations. Human computers were the people who did the math in order to write books of mathematical tables. These books of tables were essential to statisticians and astronomers and navigators and bankers and ballistics experts, all of whom relied on complex calculations for everyday use. If you needed to multiply or divide very large numbers or raise a number to the power of x or extract the nth root of a large number, it was laborious and burdensome to do such a calculation on the fly. It was easier to look up the result in a precalculated table. The system worked beautifully for years; the Egyptian mathematician Ptolemy was known to use mathematical tables in the second century AD, and in 1758, French astronomers calculated the return of Halley's Comet using only humans and mathematical lookup tables.

As the Industrial Revolution progressed, the limited supply of human computers became a significant obstacle to progress. One major vexation for nineteenth-century mathematicians was the fact that the available workforce was severely limited. Today, if you wanted to hire someone to perform calculations, you could hire across the gender spectrum. In the nineteenth century, you were limited to hiring men. Few women had enough mathematical education to perform the necessary calculations; of this small set, even fewer were supported in seeking employment outside the home. In the nineteenth century, most women weren't allowed to vote in the United States. The Seneca Falls Convention, a touchstone for the beginning of the women's rights movement, didn't happen until 1848. The Nineteenth Amendment didn't pass until 1920. Plenty of men were allies in the women's suffrage movement, but mathematicians were not known for their political activism. In *The Suffragents: How Women Used Men to Get the Vote*, my colleague Brooke Kroeger chronicles the many men who worked for women's equality. Of these men, several were professors—of history, literature, philosophy. None of them were professors of mathematics, however.[13]

The nineteenth century was also the time of America's great shame, slavery. Black men and women could have worked as human computers, could have been productive members of the workforce, except that they were enslaved as forced labor. Slaves were not allowed educational opportunities; they were beaten and raped and killed. Throughout the nineteenth century, people of color were forcibly excluded from higher education opportunities and thus from the workforce of the intellectual elites. Slavery didn't end until late in the century: Abraham Lincoln issued the Emancipation Proclamation in 1863, followed by the Thirteenth Amendment in 1865. Access to education didn't improve for decades afterward, and many people would argue that we still have far to go to provide fair, equal, and integrated education in this country.

Whether they realized it consciously or not, nineteenth-century mathematicians and other scientists had a choice. One option was to enact social change (emancipation, universal suffrage, breaking down class barriers) and develop the existing workforce by allowing all the people who weren't elite white men greater access to education and train these workers for jobs. Another option was to settle for the status quo and build machines that could do the work.

They built machines.

To be fair, they were always going to build the machines. That's where their interests lay and where the development of their field was heading, and indeed the entire world was caught up in the fervor to develop new machines that took advantage of steam power, electricity, and other marvelous advances. Perhaps it seems unfair to also expect them to be economists (no matter how closely linked the fields) and civil rights activists (a phrase that had not even been coined then). I had to use mathematical lookup tables in high school trigonometry class; it was truly onerous, and I agree completely with the labor-saving value of using machines for complex, mundane calculations. However, the importance of this history is that it demonstrates just how deep this particular strain of white, male bias in tech goes. When faced with the option of bringing more, different people into the workforce, nineteenth-century mathematicians and engineers chose instead to build machines that replaced people—at enormous profit.

Fast forward to Minsky's era, and we see how the new discipline of computer science inherited the biases of the mathematical community. As wonderfully creative as Minsky and his cohort were, they also solidified the culture of tech as a billionaire boys' club. Math, physics, and the other "hard" sciences have never been hospitable to women and people of color; tech followed this lead.

A story that physicist Stephen Wolfram tells about Minsky illuminates one of the ways the subtle assumptions about gender played out in their cohort:

The Marvin that I knew was a wonderful mixture of serious and quirky. About almost any subject he'd have something to say, most often quite unusual. Sometimes it'd be really interesting; sometimes it'd just be unusual. I'm reminded of a time in the early 1980s when I was visiting Boston and subletting an apartment from Marvin's daughter Margaret (who was in Japan at the time). Margaret had a large and elaborate collection of plants, and one day I noticed that some of them had developed nasty-looking spots on their leaves.

Being no expert on such things (and without the web to look anything up!), I called Marvin to ask what to do. What ensued was a long discussion about the possibility of developing microrobots that could chase mealybugs away. Fascinating though it was, at the end of it I still had to ask, "But what should I actually do about Margaret's plants?" Marvin replied, "Oh, I guess you'd better talk to my wife."[14]

It's a cute conversation to imagine: two preeminent scientists discussing nanobots to destroy mealybugs. However, I'm also struck by the fact that neither of them knew how to care for a houseplant. Instead, care-taking responsibilities were delegated to Minsky's wife and daughter. Both women are quite accomplished: Minsky's wife, Gloria Rudisch, was a successful pediatrician, and his daughter, Margaret, has a PhD from MIT and has run software companies. However, the women were *also* expected to know how to care for growing things, a kind of invisible labor, whereas the men weren't.

Because humans have a long and successful history of dealing with plant problems, this conversation suggests a certain learned helplessness in these scientists. It wasn't hard to diagnose houseplants "without the web" in the 1980s. You could go to the local florist with a description of the spots. You could go to the local hardware store to discuss your plant problems. You could telephone the local agricultural extension office. At any of these places, there would be a community member with the appropriate horticultural knowledge. People know how to deal with plant problems; civilization is practically synonymous with horticulture. Mealybugs can be destroyed by putting a few drops of dish soap into a spray bottle of water and squirting the infested plant. Deploying bots on houseplants is a fun idea, but it's simply unnecessary.

I get it; it's more fun to talk about wacky ideas than gender politics. This was true then and is still the case. Unfortunately, wacky ideas have dominated the public dialogue in tech to the point that important conversations about social issues have been drowned out or dismissed for years. Some of the ideas that come out of Silicon Valley include buying islands in New Zealand to prep for doomsday; *seasteading*, or building islands out of discarded shipping containers to create a new paradise without government or taxes; freezing cadavers so that the deceased's consciousness can be uploaded into a future robot body; creating oversized dirigibles; inventing a meal-replacement powder named after dystopian sci-fi movie *Soylent Green*; or making cars that fly. These ideas are certainly creative, and it's important to make space in life for dreamers—but it's equally important not to take insane ideas seriously. We should be cautious. Just because someone has made a mathematical breakthrough or made a lot of money, that doesn't mean we should listen to them when they suggest aliens are real or suggest that in the future it will be possible to reanimate people, so we should keep smart people's brains in large freezers like the ones used for frozen vegetables at Costco. (Minsky was on the scientific advisory board of Alcor Cryonics, a foundation for wealthy "transhumanist" true believers who maintain a freezer in Arizona for dead bodies and brains. The foundation's multi-million-dollar trust is designed to keep the power on for decades.)[15]

Reading about Silicon Valley billionaires' desires to live to age two hundred or talk with little green men, it's tempting to ask: Were you high when you thought of that? Often, the answer is yes. Steve Jobs dropped acid in the early 1970s after he dropped out of Reed College. Doug Engelbart, the

NASA- and ARPA-funded researcher who performed the 1968 "mother of all demos" that showed for the first time all the hardware and software elements of modern computing, dropped acid at the International Foundation for Advanced Study, the legal home for academic inquiry into LSD that lasted until 1967.

Operating the camera for Engelbart's demo was Stewart Brand, the *Whole Earth Catalog* founder who helped organize LSD guru Ken Kesey's infamous acid tests, massive drug-fueled cross-country bacchanals that were chronicled in Tom Wolfe's book *The Electric Kool-Aid Acid Test*. Brand was the most important connector between Minsky's world of scientists and the counterculture. "We are as gods and might as well get good at it," Brand wrote as the first line of the *Whole Earth Catalog* in 1968.[16] That publication was a major source of inspiration for almost all the early Internet pioneers, from Steve Jobs to tech-publishing titan Tim O'Reilly. When developers created early Internet message boards, they were trying to recreate the freewheeling commentary and recommendation culture that flourished in the back pages of the *Whole Earth Catalog*, where readers wrote in to share requests, tools, and tips on communal living. As Fred Turner writes in *From Counterculture to Cyberculture: Stewart Brand, the Whole Earth Network, and the Rise of Digital Utopianism*, Brand was everywhere in the background of early Internet development. Space colonies? Brand was speculating about them in the 1970s in his magazine *CQ*, the next iteration of the *Whole Earth Catalog* and the precursor to *Wired*, the influential technology-culture magazine that Brand also founded. Turner writes: "For the readers of *CQ*, space colonies served as a rhetorical prototype. They allowed former New Communalists to transfer their longings for a communal home to the same large-scale technologies that characterized the cold war technocracy they had sought to undermine. Fantasies of a shared, transcendent consciousness gave way to dreams of technologically enabled collaboration in friction-free space. Within a decade, these fantasies would reappear in the rhetoric of cyberspace and the electronic frontier, and as they did, they would help structure public perceptions of computer networking technology."[17]

Minsky and Brand were close friends, and Brand's book *The Media Lab* featured Minsky as a prime character. Brand's ambition, curiosity, and passion for tech fit neatly with Minsky's iconoclastic band of hackers. Looking back on the *Whole Earth* project, Brand wrote: "At a time when the New

Left was calling for grass-roots political (i.e., referred) power, Whole Earth eschewed politics and pushed grassroots direct power tools and skills. At a time when New Age hippies were deploring the intellectual world of arid abstractions, Whole Earth pushed science, intellectual endeavor, and new technology as well as old. As a result, when the most empowering tool of the century came along[,] personal computers (resisted by the New Left and despised by the New Age)[,] Whole Earth was in the thick of the development from the beginning."[18]

Brand, an Exeter and Stanford grad whose father was an MIT engineer, looked to personal computers as the new frontier for a bright, new, Utopian future.[19] He started the very first online community in 1985, the Whole Earth eLectronic Link (WELL), which is where tech developed its current political default attitude, libertarianism. Paulina Borsook chronicles the libertarian takeover of tech in *Cyberselfish: A Critical Romp through the Terribly Libertarian Culture of High Tech.* A virulent form of philosophical technolibertarianism lurks at the heart of the online communities that are most radically invested in what they call "free speech" and radical individuality. This sentiment used to thrive on message boards; in 2017, it lives on the red-pill forums of Reddit and on the dark web. Borsook writes: "It bespeaks a lack of human connection and a discomfort with the core of what many of us consider it means to be human. It's an inability to reconcile the demands of being individual with the demands of participating in society, which coincides beautifully with a preference for, and glorification of, being the solo commander of one's computer in lieu of any other economically viable behavior. Computers are so much more rule-based, controllable, fixable, and comprehensible than any human will ever be."[20] This is Turing's social awkwardness, politicized and magnified.

The transition from hippie ideology to the antigovernment ideology of cyberspace activists is visible in "A Declaration of the Independence of Cyberspace," published in 1996 by former Grateful Dead lyricist John Perry Barlow. "Governments of the Industrial World, you weary giants of flesh and steel, I come from Cyberspace, the new home of Mind," Barlow writes. "On behalf of the future, I ask you of the past to leave us alone ... You have no sovereignty where we gather. We have no elected government, nor are we likely to have one."[21] Barlow started the libertarian Electronic Frontier Foundation, which today defends hackers, because of debates he had on the WELL.

Then came Peter Thiel. Thiel, another libertarian Stanford grad—who founded PayPal, was an early investor in Facebook, and founded the CIA-backed big data firm Palantir—is frank about his hostility toward gender equality and government. In a 2009 *Cato Unbound* essay, Thiel writes, "Since 1920, the vast increase in welfare beneficiaries and the extension of the franchise to women—two constituencies that are notoriously tough for libertarians—have rendered the notion of 'capitalist democracy' into an oxymoron." Like Barlow, Thiel conceives of cyberspace as a stateless country: "Because there are no truly free places left in our world, I suspect that the mode for escape must involve some sort of new and hitherto untried process that leads us to some undiscovered country; and for this reason I have focused my efforts on new technologies that may create a new space for freedom."[22] Thiel was a supporter of and advisor to Donald Trump's presidential campaign and funded a lawsuit that took down *Gawker*. In the book *Move Fast and Break Things*, Annenberg Innovation Lab director emeritus Jonathan Taplin explores the way that Thiel's influence has spread throughout Silicon Valley via his "Paypal Mafia," other venture capitalists and executives who have adopted his anarcho-capitalist philosophy.[23]

The question of why wealthy people like Thiel are taken seriously about seasteading or aliens has been addressed by cognitive scientists. Paul Slovic, an expert in risk assessment, writes that we have cognitive fallacies related to expertise. We tend to assume that when people are experts at one thing, their expertise extends to other areas as well.[24] This is why people assume that because Turing was right about math, he was also right in his assessments of how society works. Especially in our time of highly specialized labor, this can be problematic. Being good with computers is not the same as being good with people. We shouldn't rush to be governed by computational systems designed by people who don't care about or understand the cultural systems in which we are all embedded.

The way that white male bias interacts with the genius myth inside STEM fields is even more pernicious. Even today, women and people of color are rarely considered math or tech geniuses. In 2015, Princeton professor S. J. Leslie and collaborators looked at *ability beliefs*, or how scholars prioritize genius and brilliance versus empathy and hard work in different academic fields. They write: "Across the academic spectrum, women are underrepresented in fields whose practitioners believe that raw, innate talent is the main requirement for success, because women are stereotyped as not possessing such talent. This hypothesis extends to African Americans' underrepresentation as well, as this group is subject to similar stereotypes."[25]

The negative effects of gender stereotypes associated with math are found throughout STEM disciplines. The cultures in STEM fields "impose a set of masculinized norms and expectations that limit approaches to scientific inquiry," write scholars Shane Bench, Heather Lench, and collaborators in a 2015 article. "Disciplinary norms in STEM fields dictate that scientists are decisive, methodical, objective, unemotional, competitive, and assertive—characteristics associated with men and masculinity ... Because STEM fields are stereotypically associated with men and masculinity, women perceive them as antithetical to themselves as female and that they do not belong in those contexts ... the more women perceived an environment (i.e., a computer science classroom) as masculine, the less they reported being interested in joining the field."[26]

The dynamic that Bench et al describe seems to be in effect at Minsky's alma mater, the Harvard math department. "Current and former students and faculty—male and female—say the department's dearth of female faculty and graduate students creates a discouraging environment for women undergraduates," writes Hannah Natanson in a 2017 *Harvard Crimson* article. "Women in the department are often told to take easier classes than their male peers; and, in a department dominated by men, everyday faculty-to-student and peer-to-peer interactions leave women feeling conspicuous and uncomfortable."[27] The Harvard math department does not have a single female senior faculty member. The department did appoint a woman to full professor, the highest rank in the department—but not until 2009. She left for Princeton not long afterward. Since then, three women have been offered tenured professorships. All three declined.

Bench et al. also explore how "positivity bias" contributes to the gender gap in STEM fields. In the study, they gave men and women the same math test and asked them how they thought they performed. When the researchers graded the tests and looked at the students' estimates of their scores, they found that men consistently thought they scored higher than they actually did. "This greater overestimation of performance in men accounted for their greater intent to pursue math fields compared to women," the scholars wrote. "The findings suggest that gender gaps in STEM fields are not necessarily the result of women underestimating their abilities, but rather may be due to men overestimating their abilities."

To recap: we have a small, elite group of men who tend to overestimate their mathematical abilities, who have systematically excluded women and people of color in favor of machines for centuries, who tend to want to make science fiction real, who have little regard for social convention, who don't believe that social norms or rules apply to them, who have unused piles of government money sitting around, and who have adopted the ideological rhetoric of far-right libertarian anarcho-capitalists.

What could possibly go wrong?

7 Machine Learning: The DL on ML

In order to create a more just technological world, we need more diverse voices at the table when we create technology. To do this, we need to employ conventional solutions like reducing barriers to entry and addressing the "leaky pipeline" issues that make mid-career professionals drop out or stall on their way to the top. I think we also need to add an unconventional solution: we need to add nuance to the way we talk about all things digital. This is easier said than done. One illustration of the difficulty of talking about computer science comes from an *xkcd* comic by Randall Munro. In it, a woman sits at a computer and a man stands behind her:

"When a user takes a photo, the app should check whether they're in a national park," says the man.

"Sure, easy GIS lookup," says the woman. "Gimme a few hours."

"And check whether the photo is a bird," says the man.

"I'll need a research team and five years," says the woman.

"In CS, it can be hard to explain the difference between the easy and the virtually impossible," reads the caption.[1]

Because it's complicated to explain why a computer might have trouble recognizing a bird in an image, or differentiating between a parrot and guacamole, we need more people (data journalists, perhaps?) explaining complex technical topics in plain language to demystify the more arcane corners of the AI world.

The difficulty of talking about computation has led to a lot of misunderstandings. One recurrent idea in this book is that computers are good at some things and very bad at others, and social problems arise from situations in which people misjudge how suitable a computer is for performing the task. The classic example of a thing that is very simple for people but very complex for computers is navigating a room with toys all over the floor. The average toddler can navigate a room without stepping on toys (though of course she might not choose to do so). A robot can't. To get the robot to navigate the toy-strewn floor, we would have to program in all of the information about the toys and their exact dimensions and have the robot calculate a path around the toys. If the toys moved, the robot would need its schema updated. Self-driving cars, which we'll discuss in chapter 8, work like this hypothetical robot in the playroom: they constantly update their preprogrammed map of the world.

There are also predictable pitfalls to the robot method, as people who own both Roomba robotic vacuum cleaners and pets have discovered. When a pet leaves something disgusting on the floor, the Roomba will smear it all over the house. "Quite honestly, we see this a lot," a spokesman from iRobot, the company that makes the Roomba, said to the *Guardian* in August 2016. "We generally tell people to try not to schedule your vacuum if you know you have dogs that may create such a mess. With animals anything can happen."[2]

I can use a euphemism to talk about the disgusting things that pets do because everyday language allows us to refer to things without using precise words. If I say that my dog is adorable, but also gross, you will understand. You can hold the two competing ideas in your head at the same time, and you can guess what I mean by *gross*. There are no such euphemisms in mathematical language. In mathematical language, everything is highly precise. Part of the communication problem that exists in computational culture derives from the imprecision of everyday language and the precision of mathematical language. One example: in programming, there is a concept called a *variable*. You assign a value to a variable by writing something like "X = 2," and then you can use X in a routine. There are two kinds of variables: variables that change, which are called *variables*, and variables that don't change, which are called *constants*. This makes perfect sense to a programmer: a variable can be a constant. To a nonprogrammer, it likely doesn't make sense: constant is the opposite of varying, so a thing that varies is not a thing that is unvarying. It's confusing.

This naming problem is not new. Language has always evolved along with science. In biology, cells got their name because the man who discovered them in 1665, Robert Hooke, was reminded of the walls of monks' cells in monasteries. The naming problem is particularly acute right now, however, because of the rapid pace of technological change. We're adopting new computational concepts and new hardware at a breathtaking rate, and people are inventing names for new things based on concepts or artifacts that already exist.

Although computer scientists and mathematicians tend to be talented at computer science and math, as a group they tend not to be sensitive to the nuances of language. If something needs a name, they don't obsess over picking the perfect name that has ideal connotations and Latin roots and what have you. They just pick a name, usually one that has to do with something they like. Python the programming language is named after Monty Python the comedy troupe (Monty Python is the ur-comedy text in computer science, like Star Wars is the ur-narrative text.) Django, a web framework, is named after Django Reinhardt the jazz guitarist, a favorite of the Django framework's inventor. Java the language is named after coffee. JavaScript, an unrelated language, was invented around the same time as Java and is also (unfortunately) named after coffee.

As the term *machine learning* has spread from computer science circles into the mainstream, a number of issues have arisen from linguistic confusion. Machine learning (ML) implies that the computer has agency and is somehow sentient because it "learns," and *learning* is a term usually applied to sentient beings like people (or partially sentient beings like animals). However, computer scientists know that machine "learning" is more akin to a metaphor in this case: it means that the machine can improve at its programmed, routine, automated tasks. It doesn't mean that the machine acquires knowledge or wisdom or agency, despite what the term *learning* might imply. This type of linguistic confusion is at the root of many misconceptions about computers.[3]

Imagination also complicates things. How you define AI depends on what you want to believe about the future. One of Marvin Minsky's students, Ray Kurzweil, is a proponent of the *singularity theory*, a hypothetical future merging of man and machine that he thinks will be achieved by 2045. (Kurzweil is famous for inventing a musical synthesizer that sounds like a grand piano.) Singularity is a major preoccupation of science fiction. I was once interviewed for a futurists' summit, and the interviewer asked me about the paperclip theory: What if you invented a machine that made paperclips, and then you taught the machine to want to make paperclips, and then you taught the machine to want to make other things, and then the machine made lots of other machines and all the machines took over?

"Is that the singularity?" the interviewer asked. "And aren't you worried about it?" That's fun to think about, but it's also not reasonable. You can unplug the paperclip machine. Problem solved. Also, this is a purely hypothetical situation. *It's not real.*

As psychologist Stephen Pinker told *IEEE Spectrum*, the magazine of the Institute of Electrical and Electronics Engineers (IEEE), in a special issue on the singularity: "There is not the slightest reason to believe in a coming singularity. The fact that you can visualize a future in your imagination is not evidence that it is likely or even possible. Look at domed cities, jet-pack commuting, underwater cities, mile-high buildings, and nuclear-powered automobiles—all staples of futuristic fantasies when I was a child that have never arrived. Sheer processing power is not a pixie dust that magically solves all your problems."[4]

Facebook's Yann LeCun is also a singularity skeptic. He told *IEEE Spectrum*: "There are people that you'd expect to hype the Singularity, like Ray Kurzweil. He's a futurist. He likes to have this positivist view of the future. He sells a lot of books this way. But he has not contributed anything to the science of AI, as far as I can tell. He's sold products based on technology, some of which were somewhat innovative, but nothing conceptually new. And certainly he has never written papers that taught the world anything on how to make progress in AI."[5] Reasonable, smart people disagree about what will happen in the future—in part because nobody can see the future.

I'm going to try to bring some clarity to the situation by defining machine learning and showing you an example of how someone might perform machine learning on a dataset. I'm going to explain machine learning a few different ways and also demonstrate some code. It's going to get technical. If the technical parts get confusing, don't worry; you can skim them first and return to them later.

AI enjoyed a popularity bump in 2017 in contrast to many years of what people call an *AI winter*. In the mainstream, people mostly ignored AI for the first decade of the 2000s. The Internet was the popular thing technologically, then mobile devices, and those were the focus of our collective imagination. In the middle of the 2010s, however, people started talking about machine learning. Suddenly, AI was on fire again. AI startups were founded and acquired. IBM's Watson beat a human player at Jeopardy!; an algorithm outfoxed a human player at playing Go. Even the words *machine learning* were cool. A machine could learn! The promise was delivered!

At first, I wanted to believe that some genius had figured out the truly hard problem of making a machine think—but when I looked closer, it turned out that the reality was far more nuanced. What had happened was that scientists had redefined the term *machine learning* so that it referred to their work. They used the term so much that its meaning changed.

This happens. Language is fluid. A good example is the word *literally*, which used to mean the opposite of *figuratively*. In the 1990s, if you said, "My mouth was literally on fire after eating that ghost pepper," it meant that there were actual flames in your mouth and you were talking from the other side of recovery from third-degree burns. However, in the 2000s, a critical mass of people started using *literally* as a synonym for figuratively and for emphasis. "I was ready to literally kill someone if I had to listen to that John Mayer song one more time" became understood as "I would really prefer not to listen to another John Mayer song," rather than a statement about murder or mayhem.

The term *machine learning* entered the lexicon in 1959, according to the *Oxford English Dictionary* (OED). The OED began including *machine learning* as a phrase in its third edition, published in 2000. The OED defines machine learning as follows:

machine learning   n. Computing the capacity of a computer to learn from experience, i.e. to modify its processing on the basis of newly acquired information.

1959   IBM Jrnl. 3 211/1   We have at our command computers with adequate data-handling ability and with sufficient computational speed to make use of machine-learning techniques.

1990   New Scientist 8 Sept. 78/1   When Doug Lenat of Stanford developed Eurisko, a second generation machine learning system, he thought that he had created a real intellectual.[6]

This definition is true, but it doesn't quite capture the way that contemporary computer scientists use the term. A more comprehensive definition is found in Oxford's *A Dictionary of Computer Science*:

machine learning

A branch of artificial intelligence concerned with the construction of programs that learn from experience. Learning may take many forms, ranging from learning from examples and learning by analogy to autonomous learning of concepts and learning by discovery.

Incremental learning involves continuous improvement as new data arrives while *one-shot* or *batch learning* distinguishes a training phase from the application phase.

Supervised learning occurs when the training input has been explicitly labeled with the classes to be learned.

Most learning methods aim to demonstrate generalization whereby the system develops efficient and effective representations that encompass large chunks of closely related data.[7]

This is closer, but still not quite right. The documentation for scikit-learn, a popular software library for machine learning in Python, has a different definition: "Machine learning is about learning some properties of a data set and applying them to new data. This is why a common practice in machine learning to evaluate an algorithm is to split the data at hand into two sets, one that we call the training set on which we learn data properties and one that we call the testing set on which we test these properties."[8]

It's rare that a term has so much disagreement across different sources. The definition of a dog, for example, is pretty consistent across texts. However, machine learning is so new, and there is so little consensus, that it's not surprising that the linguistic definitions haven't caught up to reality.

Tom M. Mitchell, the E. Fredkin University Professor in the Machine Learning Department of Carnegie Mellon University's School of Computer Science, offers a good definition of machine learning in "The Discipline of Machine Learning." He writes: "We say that a machine learns with respect to a particular task T, performance metric P, and type of experience E, if the system reliably improves its performance P at task T, following experience E. Depending on how we specify T, P, and E, the learning task might also be called by names such as data mining, autonomous discovery, database updating, programming by example, etc."[9] I think this is a good definition because Mitchell uses very precise language to define learning. When a machine "learns," it doesn't mean that the machine has a brain made out of metal. It means that the machine has become more accurate at performing a single, specific task according to a specific metric that a person has defined.

This kind of learning does *not* imply intelligence. As programmer and consultant George V. Neville-Neil writes in the *Communications of the ACM*:

We have had nearly 50 years of human/computer competition in the game of chess, but does this mean that any of those computers are intelligent? No, it does not—for two reasons. The first is that chess is not a test of intelligence; it is the test of a particular skill—the skill of playing chess. If I could beat a Grandmaster at chess and yet not be able to hand you the salt at the table when asked, would I be intelligent? The second reason is that thinking chess was a test of intelligence was based on a false cultural premise that brilliant chess players were brilliant minds, more gifted than those around them. Yes, many intelligent people excel at chess, but chess, or any other single skill, does not denote intelligence.[10]

There are three general types of machine learning: supervised learning, unsupervised learning, and reinforcement learning. Here are definitions of each from a widely used textbook called *Artificial Intelligence: A Modern Approach* by UC Berkeley professor Stuart Russell and Google's director of research, Peter Norvig:

Supervised learning: The computer is presented with example inputs and their desired outputs, given by a "teacher," and the goal is to learn a general rule that maps inputs to outputs.

Unsupervised learning: No labels are given to the learning algorithm, leaving it on its own to find structure in its input. Unsupervised learning can be a goal in itself (discovering hidden patterns in data) or a means toward an end (feature learning).

Reinforcement learning: A computer program interacts with a dynamic environment in which it must perform a certain goal (such as driving a vehicle or playing a game against an opponent). The program is provided feedback in terms of rewards and punishments as it navigates its problem space.[11]

Supervised learning is the most straightforward. The machine is provided with the training data and labeled outputs. We essentially tell the machine what we want to find, then fine-tune the model until we get the machine to predict what we know to be true.

All three kinds of machine learning depend on *training data*, known datasets for practicing and tuning the machine-learning model. Let's say that my training data is a dataset of one hundred thousand credit card company customers. The dataset contains the data you would expect a credit card company to have for a person: name, age, address, credit score, interest rate, account balance, name(s) of any joint signers on the account, a list of charges, and a record of payment amounts and dates. Let's say that we want the ML model to predict who is likely to pay their bill late. We want to find these people because every time someone pays a bill late, the interest rate on the account increases, which means the credit card company makes more money on interest charges. The training data has a column that indicates who in this group of one hundred thousand *has* paid their bills late. We split the training data into two groups of fifty thousand names each: the training set and the test data. Then, we run a machine-learning algorithm against the training set to construct a model, a black box, that predicts what we already know. We can then apply the model to the test data and see the model's prediction for which customers are likely to pay late. Finally, we compare the model's prediction to what we know is true—the customers in the test data who actually paid late. This gives us a score that measures the model's precision and recall. If we as model makers decide that the model's precision/recall score is high enough, we can deploy the model on real customers.

A handful of different machine-learning algorithms are available to apply to datasets. You may have come across some of the names, which include random forest, decision tree, nearest neighbor, naive Bayes, or hidden Markov. An *algorithm*, remember, is a series of steps or procedures that the computer is instructed to follow. In machine learning, the algorithm is coupled with variables to create a mathematical model. A wonderful explanation of models is found in Cathy O'Neil's *Weapons of Math Destruction*. O'Neil explains that we model things unconsciously all the time. When I decide what to make for dinner, I make a model: what food is in my refrigerator, what dishes I could possibly make with that food, who the people eating that night are (usually my husband and son and me), and what their food preferences are. I evaluate the various dishes and recall how each performed in the past—who took seconds of what, and what items are on the ever-changing list of shunned foods: cashews, frozen vegetables, coconut, organ meats. By deciding what to make based on what I have and what people like, I'm optimizing my meal choices for a set of features. Building a mathematical model means formalizing the features and the choices in mathematical terms.[12]

Let's say that I want to "do" machine learning. The first thing I do is grab a dataset. A variety of interesting datasets are available for machine-learning practice; they are collected in online repositories. There are datasets of facial expressions, of pets, or of YouTube videos. There are datasets of emails sent by people who worked at a failed company (Enron), datasets of newsgroup conversations in the 1990s (Usenet), datasets of friendship networks from failed social network companies (Friendster), datasets of movies that people watched on streaming services (Netflix), datasets of people saying common phrases in different accents, or datasets of people's messy handwriting. These datasets are collected from active corporations, from websites, from university researchers, from volunteers, and from defunct corporations. This small number of iconic datasets is posted online and the datasets form the backbone of all contemporary artificial intelligence. You might even find your own data in them. A friend of mine once found a video of herself as a toddler in a behavioral science archive; her mother had participated in a parent-child behavioral study when my friend was little. Researchers still had the video and still used it for drawing conclusions about the world.

Now, let's go through a classic practice exercise: we'll use machine learning to predict who survived the *Titanic* crash. Think about what happened on the *Titanic* after it hit the iceberg. Did you picture Leonardo di Caprio and Kate Winslet sliding across the decks of the ship? That's not real—but it probably colors your recall of the event, if you've seen the movie as many times as I have. It's quite likely that you've seen the movie at least once. *Titanic* earned $659 million and $1.5 billion overseas, making it the biggest movie in the world in 1997 and the second-highest-grossing film ever worldwide. (*Titanic* director James Cameron also holds the number-one spot for his other blockbuster, *Avatar*.) The film stayed in theaters for almost a year, fueled in part by young people who went to the theater to watch it over and over again.[13] *Titanic* the movie has become a part of our collective memory, just like the actual *Titanic* maritime disaster. Our brains quite commonly confuse actual events with realistic fiction. It's unfortunate, but it's normal. This confusion complicates the way we perceive risk.

We draw conclusions about risk based on *heuristics*, or informal rules. These heuristics are affected by stories that are easy to recall and by emotionally resonant experiences. For example: When he was a little boy, *New York Times* columnist Charles Blow was attacked by a vicious dog. The dog almost tore his face off. As an adult, he writes in his memoir, he remains wary of strange dogs.[14] This makes perfect sense. Being a small child attacked by a large animal is traumatic, and of course it would be the first thing someone would think of when seeing a dog for the rest of his life. Reading the book, I empathized with the little boy and felt scared when he felt scared. The day after I read Blow's memoir, I saw a man walking a dog without a leash in a park near my house—and I immediately thought of Blow and how other people who are afraid of dogs would be made uncomfortable by the fact that this dog was not on a leash. I wondered if the dog would go berserk and, if so, what would happen. The story affected my perception of risk. This is the same thinking that leads people to carry pepper spray after watching a lot of episodes of *Law & Order: SVU* or to check the back seat of the car for nasty surprises after watching a horror movie. The technical name is the *availability heuristic*.[15] The stories that spring to mind first are the ones we tend to think are the most important or occur most frequently.

Perhaps because it features so prominently in our collective imagination, the *Titanic* disaster is commonly used for teaching machine learning. Specifically, a list of the passengers on the *Titanic* is used to teach students how to generate predictions using data. It works well as a class exercise because almost all of the students have seen *Titanic* or know about the disaster. This is valuable for an instructor because you don't have to spend too much class time going over the historical context: you can get right to the fun part, which is the prediction.

I'm going to take you through the fun part using supervised learning. I think it is important to see exactly what happens when someone does machine learning. There are plenty of sites online that have ML tutorials if you're interested in going through the exercise yourself. I'm going to take you through a tutorial from a site called DataCamp, which was recommended as a first step for competing in data-science competitions by a different site, Kaggle.[16] Kaggle, which is owned by Google's parent company, Alphabet, is a site in which people compete to get the highest score for analyzing a dataset. Data scientists use it to compete in teams, sharpen their skills, or practice collaborating. It's also useful for teaching students about data science or for finding datasets.

We're going to do a DataCamp *Titanic* tutorial using Python and a few popular Python libraries: pandas, scikit-learn, and numpy. A *library* is a little bucket of functions sitting somewhere on the Internet. When we import a library, we make its functions available to the program we're writing. One way to think about it is to think about a physical library. I'm a member of the New York Public Library (NYPL). Whenever I go to stay somewhere for more than a week, for work or for vacation, I generally try to go to the local library and get a library card. Signing up for a local library card allows me to use all the books and resources available at that library. For the time that I'm a local library member, I can use all my core NYPL resources *plus* the unique resources of the local library. In a Python program, we start with a whole bunch of built-in functions: those are the NYPL. Importing a new library is like signing up for the local library card. Our program can use all the good stuff in the core Python library *plus* the nifty functions written by the researchers and open-source developers who made and published the scikit-learn library, for example.

Pandas, another library we'll use, has a container called a *DataFrame* that "holds" a set of data. This type of container is also called an *object*, as in *object-oriented programming*. *Object* is a generic term in programming, just as it is in the real world. In programming, an *object* is a conceptual wrapper for a little package of data, variables, and code. Having the label *object* gives us something to hold on to. We need to conceptualize our package of bits as something in order to think about it and talk about it.

The first thing we do is break our data into two sets: training data and test data. We're going to develop a model, train it on the training data, then run it on the test data. Remember how there is general AI and narrow AI? This is narrow. Let's start by typing the following:

```
import pandas as pd
import numpy as np
from sklearn import tree, preprocessing
```

We've just imported several libraries that we'll use for our analysis. We use an alias, *pd*, for pandas, and the alias *np* for numpy. We now have access to all of the functions in pandas and numpy. We can choose to import all of the functions or just a few. From scikit-learn, we'll import only two functions. One is named *tree* and the other is named *preprocessing*.

Next, let's import the data from a comma-separated values (CSV) file that is also sitting somewhere on the Internet. Specifically, this CSV file is sitting on a server owned by Amazon Web Services (AWS). We can tell because the base URL of the file (the first part after http://) is s3.amazonaws.com. A CSV file is a file of structured data in which each column is separated by a comma. We're going to import two different Titanic data files from AWS. One is a training data set, another is a test data set. Both data sets are in CSV format. Let's import the data:

```
train_url =
"http://s3.amazonaws.com/assets.datacamp.com/course/Kaggle/
train.csv"
train = pd.read_csv(train_url)

test_url = "http://s3.amazonaws.com/assets.datacamp.com/
course/Kaggle/test.csv"
test = pd.read_csv(test_url)
```

pd.read_csv() means "please invoke the read_csv() function, which lives in the pd (pandas) library." Technically, we created a DataFrame object and called one of its built-in methods. Regardless, the data is now imported into two variables: *train* and *test*. We'll use the data in the *train* variable to create the model, and then we'll use the data in the *test* variable to test our model's accuracy.

Let's see what's in the *head*, or the first few lines, of the training data:

```
print(train.head())
```

	PassengerId	Survived	Pclass \
0	1	0	3
1	2	1	1
2	3	1	3
3	4	1	1
4	5	0	3

	Name	Sex	Age	SibSp \
0	Braund, Mr. Owen Harris	male	22.0	1
1	Cumings, Mrs. John Bradley (Florence Briggs Th...	female	38.0	1
2	Heikkinen, Miss. Laina	female	26.0	0
3	Futrelle, Mrs. Jacques Heath (Lily May Peel)	female	35.0	1
4	Allen, Mr. William Henry	male	35.0	0

	Parch	Ticket	Fare	Cabin	Embarked
0	0	A/5 21171	7.2500	NaN	S
1	0	PC 17599	71.2833	C85	C
2	0	STON/O2. 3101282	7.9250	NaN	S
3	0	113803	53.1000	C123	S
4	0	373450	8.0500	NaN	S

It looks like the data is twelve columns. The columns are labeled PassengerId, Survived, Pclass, Name, Sex, Age, SibSp, Parch, Ticket, Fare, Cabin, and Embarked. What do these column headings mean?

To answer this, we need a data dictionary, which is provided with most datasets. The data dictionary reveals the following:

```
Pclass = Passenger Class (1 = 1st; 2 = 2nd; 3 = 3rd)
Survived = Survival (0 = No; 1 = Yes)
Name = Name
Sex = Sex
```

```
Age = Age (in years; fractional if age less than one (1). If
the age is estimated, it is in the form xx.5)
Sibsp = Number of Siblings/Spouses Aboard
Parch = Number of Parents/Children Aboard
Ticket = Ticket Number
Fare = Passenger Fare (pre-1970 British pound)
Cabin = Cabin number
Embarked = Port of Embarkation (C = Cherbourg; Q = Queenstown;
S = Southampton)
```

For most of the columns, we have data. For some column values, we do not have data. For PassengerId 1, Mr. Owen Harris Braund, the value for Cabin is NaN. This means "not a number." NaN is different than zero; zero is a number. NaN means that there is no value for this variable. This distinction might seem unimportant for everyday life, but it's crucially important in computer science. Remember that mathematical language is precise. For example, NULL indicates an empty set, which is also different than NaN or zero.

Let's see what's in the first few lines of the test dataset:

```
print(test.head())
```

	PassengerId	Pclass	Name	Sex \
0	892	3	Kelly, Mr. James	male
1	893	3	Wilkes, Mrs. James (Ellen Needs)	female
2	894	2	Myles, Mr. Thomas Francis	male
3	895	3	Wirz, Mr. Albert	male
4	896	3	Hirvonen, Mrs. Alexander (Helga E Lindqvist)	female

	Age	SibSp	Parch	Ticket	Fare	Cabin	Embarked
0	34.5	0	0	330911	7.8292	NaN	Q
1	47.0	1	0	363272	7.0000	NaN	S
2	62.0	0	0	240276	9.6875	NaN	Q
3	27.0	0	0	315154	8.6625	NaN	S
4	22.0	1	1	3101298	12.2875	NaN	S

As you can see, *test* has the same type of data as *train*, minus the Survived column. Great! Our goal is to create a Survived column in the *test* data that contains a prediction for each passenger. (Of course, someone already knows which passengers in the *test* data set survived—but it wouldn't be much of a tutorial if the data set already contained the answers.)

Next, we're going to run some basic summary statistics on the training dataset in order to get to know it a little better. When data journalists do this, we call it *interviewing the data*. We interview data just like we might interview a human source. A human has a name, an age, a background; a dataset has a size and a number of columns. Asking a column of data about its average value is a bit like asking someone to spell their last name.

We can get to know our data a bit by running a function called *describe* that assembles some basic summary statistics and puts them into a handy table, as follows:

```
train.describe()
```

	PassengerId	Survived	Pclass	Age	SibSp	Parch	Fare
count	891.000000	891.000000	891.000000	714.000000	891.000000	891.000000	891.000000
mean	446.000000	0.383838	2.308642	29.699118	0.523008	0.381594	32.204208
std	257.353842	0.486592	0.836071	14.526497	1.102743	0.806057	49.693429
min	1.000000	0.000000	1.000000	0.420000	0.000000	0.000000	0.000000
25%	223.500000	0.000000	2.000000	20.125000	0.000000	0.000000	7.910400
50%	446.000000	0.000000	3.000000	28.000000	0.000000	0.000000	14.454200
75%	668.500000	1.000000	3.000000	38.000000	1.000000	0.000000	31.000000
max	891.000000	1.000000	3.000000	80.000000	8.000000	6.000000	512.329200

The training dataset has 891 records. Of these, only 714 records show the age of the passenger. For the data we have available, the average age of the passengers is 29.699118; normal people would say that the average age is thirty.

A few of these statistics require interpretation: Survived has a min of 0 and a max of 1. In other words, it is a Boolean value. Either someone survived (1), or they didn't (0). We can calculate an average, which turns out to be 0.38. Similarly, we can calculate an average for Pclass, or passenger class. Passengers' tickets were for first, second, or third class. The average doesn't literally mean that someone traveled 2.308 class.

Now that we've gotten to know our data a little bit, it's time to do some analysis. Let's first look at the number of passengers. We can use a function called *value_counts* to do this. Value_counts will show how many values there are for each distinct category in a column. In other words, how many passengers are traveling in each passenger class? Let's find out:

```
train["Pclass"].value_counts()
1    216
2    184
3    491
Name: Pclass, dtype: int64
```

The training data shows 491 passengers traveling third class, 184 passengers traveling second class, and 216 passengers traveling first class.

Let's look at the numbers for survival:

```
train["Survived"].value_counts()
0    549
1    342
Name: Survived, dtype: int64
```

The training data shows that 549 people perished and 342 survived.

Let's see those numbers normalized:

```
print(train["Survived"].value_counts(normalize = True))
0    0.616162
1    0.383838
Name: Survived, dtype: float64
```

Sixty-two percent of passengers perished, and 38 percent survived. In other words, most people died in the disaster. If we were to make a prediction about whether a random passenger survived, we'd likely predict that they did not survive.

We could stop here if we wanted. We just drew a conclusion that would allow us to make a reasonable prediction. We can do better, however, so let's keep going. Are there any factors that might help improve the prediction? In addition to survival, we have some other columns in the data: Pclass, Name, Sex, Age, SibSp, Parch, Ticket, Fare, Cabin, and Embarked.

Pclass is a proxy for the socioeconomic class of the passengers. That might be useful as a predictor. We could guess that first-class passengers got on the boats before third-class passengers. Sex is also a reasonable guess for a predictor. We know that "women and children first" was a principle used during maritime disasters. This principle dates to 1852, when the British HMS *Birkenhead*, a troop ship, ran aground off the coast of South Africa. It's not a uniformly applied principle, but it's recurrent enough to use for social analysis.

Now, let's do some comparisons to see if we can find variables that seem predictive:

```
# Passengers that survived vs passengers that passed away
print(train["Survived"].value_counts())
0    549
1    342
Name: Survived, dtype: int64
```

```
# As proportions
print(train["Survived"].value_counts(normalize = True))
0    0.616162
1    0.383838
Name: Survived, dtype: float64
```

```
# Males that survived vs males that passed away
print(train["Survived"][train["Sex"] == 'male'].value_counts())
0    468
1    109
Name: Survived, dtype: int64
```

```
# Females that survived vs females that passed away
print(train["Survived"][train["Sex"] == 'female'].value_counts())
1    233
0     81
Name: Survived, dtype: int64
```

```
# Normalized male survival
print(train["Survived"][train["Sex"] == 'male'].value_counts
(normalize=True))
0    0.811092
1    0.188908
Name: Survived, dtype: float64
```

```
# Normalized female survival
print(train["Survived"][train["Sex"] == 'female'].value_counts
(normalize=True))
1    0.742038
0    0.257962
Name: Survived, dtype: float64
```

We can see that 74 percent of females survived, and only 18 percent of males survived. Therefore, for a random person, we might adjust our guess to say that they survived if they were female, but not if they were male.

Remember that the goal at the beginning of this section was to create a Survived column in the test data that contains a prediction for each passenger. At this point, we could create a Survived column and fill in "1" (meaning "yes, this passenger survived") for 74 percent of the females and "0" (meaning "no, this passenger did not survive") for the remaining females. We could fill in "1" for 18 percent of the male passengers and "0" for 81 percent of the remaining males.

But we won't, because that would mean assigning probable outcomes randomly based only on gender. We know there are other factors in the data that influence the outcome. (If you're truly curious to see the nitty-gritty of how this is determined, I encourage you to look at the DataCamp tutorial or something similar online.) What about women traveling third class? Women traveling first class? Women traveling with spouses? Women traveling with children? This quickly becomes tedious to calculate manually, so we're going to train a model to do the guessing for us based on the factors that we know.

To construct the model, we're going to use a *decision tree*, a type of algorithm. Remember, there are a handful of algorithms that are standard in machine learning. They have names like decision tree, or random forest, or artificial neural network, or naive Bayes, or k-nearest neighbor, or deep learning. Wikipedia's list of machine-learning algorithms is quite comprehensive.

These algorithms come packaged into software like pandas. Very few people write their own algorithms for machine learning; it's much easier to use one that already exists. Writing a new algorithm is like writing a new programming language. You really have to care *a lot* and you have to devote a lot of time to doing it. I'm going to wave my hands and say "math" to explain what happens inside the model. Sorry. If you really want to know, I encourage you to read more about it. It's very interesting, but it's beyond the scope of the current discussion.

Now, let's train the model on the training data. We know from our exploratory analysis that the features that matter are fare class and sex. We want to create a guess for survival. We already know whether the passengers in the training data survived or not. We're going to make the model guess, then compare the guesses to reality. Whatever the percentage is that we get right is our accuracy number.

Here's an open secret of the big data world: *all data is dirty*. All of it. Data is made by people going around and counting things or made by sensors that are made by people. In every seemingly orderly column of numbers, there is noise. There is mess. There is incompleteness. This is life. The problem is, dirty data doesn't compute. Therefore, in machine learning, sometimes we have to make things up to make the functions run smoothly.

Are you horrified yet? I was, the first time I realized this. As a journalist, I don't get to make anything up. I need to fact-check each line and justify it for a fact-checker or an editor or my audience—but in machine learning, people often make stuff up when it's convenient.

Now, in physics you can do this. If you want to find the temperature at point A inside a closed container, you take the temperature at two other equidistant points (B and C) and assume that the temperature at point A is halfway between the B and C temperatures. In statistics ... well, this is how it works, and the *missing-ness* contributes to the inherent uncertainty of the whole endeavor. We'll use a function called *fillna* to fill in all of the missing values:

```
train["Age"] = train["Age"].fillna(train["Age"].median())
```

The algorithm can't run with missing values. Thus, we need to make up the missing values. Here, DataCamp recommends using the median.

Let's take a look at the data to see what's in there:

```
# Print the train data to see the available features
print(train)
```

	PassengerId	Survived	Pclass \
0	1	0	3
1	2	1	1
2	3	1	3
3	4	1	1
4	5	0	3
5	6	0	3
6	7	0	1
7	8	0	3
8	9	1	3
9	10	1	2
10	11	1	3
11	12	1	1
12	13	0	3
13	14	0	3
14	15	0	3
15	16	1	2
16	17	0	3

	PassengerId	Survived	Pclass \
17	18	1	2
18	19	0	3
19	20	1	3
20	21	0	2
21	22	1	2
22	23	1	3
23	24	1	1
24	25	0	3
25	26	1	3
26	27	0	3
27	28	0	1
28	29	1	3
29	30	0	3
..	...	...	...
861	862	0	2
862	863	1	1
863	864	0	3
864	865	0	2
865	866	1	2
866	867	1	2
867	868	0	1
868	869	0	3
869	870	1	3
870	871	0	3
871	872	1	1
872	873	0	1
873	874	0	3
874	875	1	2
875	876	1	3
876	877	0	3
877	878	0	3
878	879	0	3
879	880	1	1
880	881	1	2
881	882	0	3
882	883	0	3
883	884	0	2
884	885	0	3
885	886	0	3
886	887	0	2
887	888	1	1

	PassengerId	Survived	Pclass \
888	889	0	3
889	890	1	1
890	891	0	3

	Name	Sex	Age	SibSp \
0	Braund, Mr. Owen Harris	male	22.0	1
1	Cumings, Mrs. John Bradley (Florence Briggs Th...	female	38.0	1
2	Heikkinen, Miss. Laina	female	26.0	0
3	Futrelle, Mrs. Jacques Heath (Lily May Peel)	female	35.0	1
4	Allen, Mr. William Henry	male	35.0	0
5	Moran, Mr. James	male	28.0	0
6	McCarthy, Mr. Timothy J	male	54.0	0
7	Palsson, Master. Gosta Leonard	male	2.0	3
8	Johnson, Mrs. Oscar W (Elisabeth Vilhelmina Berg)	female	27.0	0
9	Nasser, Mrs. Nicholas (Adele Achem)	female	14.0	1
10	Sandstrom, Miss. Marguerite Rut	female	4.0	1
11	Bonnell, Miss. Elizabeth	female	58.0	0
12	Saundercock, Mr. William Henry	male	20.0	0
13	Andersson, Mr. Anders Johan	male	39.0	1
14	Vestrom, Miss. Hulda Amanda Adolfina	female	14.0	0
15	Hewlett, Mrs. (Mary D Kingcome)	female	55.0	0
16	Rice, Master. Eugene	male	2.0	4
17	Williams, Mr. Charles Eugene	male	28.0	0
18	Vander Planke, Mrs. Julius (Emelia Maria Vande...	female	31.0	1
19	Masselmani, Mrs. Fatima	female	28.0	0
20	Fynney, Mr. Joseph J	male	35.0	0
21	Beesley, Mr. Lawrence	male	34.0	0
22	McGowan, Miss. Anna "Annie"	female	15.0	0
23	Sloper, Mr. William Thompson	male	28.0	0
24	Palsson, Miss. Torborg Danira	female	8.0	3
25	Asplund, Mrs. Carl Oscar (Selma Augusta Emilia...	female	38.0	1
26	Emir, Mr. Farred Chehab	male	28.0	0
27	Fortune, Mr. Charles Alexander	male	19.0	3
28	O'Dwyer, Miss. Ellen "Nellie"	female	28.0	0
29	Todoroff, Mr. Lalio	male	28.0	0
..	...	...	...	...

	Name	Sex	Age	SibSp \
861	Giles, Mr. Frederick Edward	male	21.0	1
862	Swift, Mrs. Frederick Joel (Margaret Welles Ba...	female	48.0	0
863	Sage, Miss. Dorothy Edith "Dolly"	female	28.0	8
864	Gill, Mr. John William	male	24.0	0
865	Bystrom, Mrs. (Karolina)	female	42.0	0
866	Duran y More, Miss. Asuncion	female	27.0	1
867	Roebling, Mr. Washington Augustus II	male	31.0	0
868	van Melkebeke, Mr. Philemon	male	28.0	0
869	Johnson, Master. Harold Theodor	male	4.0	1
870	Balkic, Mr. Cerin	male	26.0	0
871	Beckwith, Mrs. Richard Leonard (Sallie Monypeny)	female	47.0	1
872	Carlsson, Mr. Frans Olof	male	33.0	0
873	Vander Cruyssen, Mr. Victor	male	47.0	0
874	Abelson, Mrs. Samuel (Hannah Wizosky)	female	28.0	1
875	Najib, Miss. Adele Kiamie "Jane"	female	15.0	0
876	Gustafsson, Mr. Alfred Ossian	male	20.0	0
877	Petroff, Mr. Nedelio	male	19.0	0
878	Laleff, Mr. Kristo	male	28.0	0
879	Potter, Mrs. Thomas Jr (Lily Alexenia Wilson)	female	56.0	0
880	Shelley, Mrs. William (Imanita Parrish Hall)	female	25.0	0
881	Markun, Mr. Johann	male	33.0	0
882	Dahlberg, Miss. Gerda Ulrika	female	22.0	0
883	Banfield, Mr. Frederick James	male	28.0	0
884	Sutehall, Mr. Henry Jr	male	25.0	0
885	Rice, Mrs. William (Margaret Norton)	female	39.0	0
886	Montvila, Rev. Juozas	male	27.0	0
887	Graham, Miss. Margaret Edith	female	19.0	0
888	Johnston, Miss. Catherine Helen "Carrie"	female	28.0	1
889	Behr, Mr. Karl Howell	male	26.0	0
890	Dooley, Mr. Patrick	male	32.0	0

	Parch	Ticket	Fare	Cabin	Embarked
0	0	A/5 21171	7.2500	NaN	S
1	0	PC 17599	71.2833	C85	C
2	0	STON/O2. 3101282	7.9250	NaN	S
3	0	113803	53.1000	C123	S

	Parch	Ticket	Fare	Cabin	Embarked
4	0	373450	8.0500	NaN	S
5	0	330877	8.4583	NaN	Q
6	0	17463	51.8625	E46	S
7	1	349909	21.0750	NaN	S
8	2	347742	11.1333	NaN	S
9	0	237736	30.0708	NaN	C
10	1	PP 9549	16.7000	G6	S
11	0	113783	26.5500	C103	S
12	0	A/5. 2151	8.0500	NaN	S
13	5	347082	31.2750	NaN	S
14	0	350406	7.8542	NaN	S
15	0	248706	16.0000	NaN	S
16	1	382652	29.1250	NaN	Q
17	0	244373	13.0000	NaN	S
18	0	345763	18.0000	NaN	S
19	0	2649	7.2250	NaN	C
20	0	239865	26.0000	NaN	S
21	0	248698	13.0000	D56	S
22	0	330923	8.0292	NaN	Q
23	0	113788	35.5000	A6	S
24	1	349909	21.0750	NaN	S
25	5	347077	31.3875	NaN	S
26	0	2631	7.2250	NaN	C
27	2	19950	263.0000	C23 C25 C27	S
28	0	330959	7.8792	NaN	Q
29	0	349216	7.8958	NaN	S
..	...	...	...	...	...
861	0	28134	11.5000	NaN	S
862	0	17466	25.9292	D17	S
863	2	CA. 2343	69.5500	NaN	S
864	0	233866	13.0000	NaN	S
865	0	236852	13.0000	NaN	S
866	0	SC/PARIS 2149	13.8583	NaN	C
867	0	PC 17590	50.4958	A24	S
868	0	345777	9.5000	NaN	S
869	1	347742	11.1333	NaN	S
870	0	349248	7.8958	NaN	S
871	1	11751	52.5542	D35	S
872	0	695	5.0000	B51 B53 B55	S
873	0	345765	9.0000	NaN	S
874	0	P/PP 3381	24.0000	NaN	C

	Parch	Ticket	Fare	Cabin	Embarked
875	0	2667	7.2250	NaN	C
876	0	7534	9.8458	NaN	S
877	0	349212	7.8958	NaN	S
878	0	349217	7.8958	NaN	S
879	1	11767	83.1583	C50	C
880	1	230433	26.0000	NaN	S
881	0	349257	7.8958	NaN	S
882	0	7552	10.5167	NaN	S
883	0	C.A./SOTON 34068	10.5000	NaN	S
884	0	SOTON/OQ 392076	7.0500	NaN	S
885	5	382652	29.1250	NaN	Q
886	0	211536	13.0000	NaN	S
887	0	112053	30.0000	B42	S
888	2	W./C. 6607	23.4500	NaN	S
889	0	111369	30.0000	C148	C
890	0	370376	7.7500	NaN	Q

[891 rows x 12 columns]

If you read all of those hundreds of lines, bravo—but if you skipped ahead, I'm not surprised. I printed many rows of data here, instead of using a small subset, in order to illustrate what it feels like to be a data scientist. Working with columns of numbers feels value-neutral and occasionally tedious. There's a certain amount of dehumanization that occurs when you deal only with numbers. It's not easy to remember that each row in a dataset represents a real person with hopes, dreams, a family, and a history.

Now that we've looked at the raw data, it's time to start working with it. Let's turn it into *arrays*, which are structures that the computer can manipulate:

```
# Create the target and features numpy arrays: target,
features_one
target = train["Survived"].values

# Preprocess
encoded_sex = preprocessing.LabelEncoder()

# Convert into numbers
train.Sex = encoded_sex.fit_transform(train.Sex)
features_one = train[["Pclass," "Sex," "Age," "Fare"]].values
```

```
# Fit the first decision tree: my_tree_one
my_tree_one = tree.DecisionTreeClassifier()
my_tree_one = my_tree_one.fit(features_one, target)
```

What we're doing is running a function called *fit* on the decision tree classifier called *my_tree_one*. The features we want to consider are Pclass, Sex, Age, and Fare. We're instructing the algorithm to figure out what relationship among these four predicts the value in the target field, which is Survived:

```
# Look at the importance and score of the included features
print(my_tree_one.feature_importances_)
[ 0.12315342  0.31274009  0.22675108  0.3373554 ]
```

The feature_importances attribute shows the statistical significance of each predictor.

The largest number in this group of values is the considered the most important:

```
Pclass = 0.1269655
Sex = 0.31274009
Age = 0.23914906
Fare = 0.32114535
```

Fare is the largest number. We can conclude that passenger fare is the most important factor in determining whether a passenger survived the *Titanic* disaster.

At this point in our data analysis, we can run a function to show exactly how accurate our calculation is within the mathematical constraints of the universe represented by this data. Let's use the score function to find the mean accuracy:

```
print(my_tree_one.score(features_one, target))
0.977553310887
```

Wow, 97 percent! That feels great. If I got a 97 percent on an exam, I'd be perfectly content. We could call this model 97 percent accurate. The machine just "learned" in that it constructed a mathematical model. The model is stored in the object called *my_tree_one*.

Next, we'll take this model and apply it to the set of test data. Remember: the test data doesn't have a Survived column. Our job is to use the model to predict whether each passenger in the test data survived or perished. We know that fare is the most important predictor according to this model, but

age and sex and passenger class matter mathematically too. Let's apply the
model to the test data and see what happens:

```python
# Fill any missing fare values with the median fare
test["Fare"] = test["Fare"].fillna(test["Fare"].median())

# Fill any missing age values with the median age
test["Age"] = test["Age"].fillna(test["Age"].median())

# Preprocess
test_encoded_sex = preprocessing.LabelEncoder()
test.Sex = test_encoded_sex.fit_transform(test.Sex)

# Extract important features from the test set: Pclass, Sex,
Age, and Fare
test_features = test[["Pclass," "Sex," "Age," "Fare"]].values
print('These are the features:\n')
print(test_features)

# Make a prediction using the test set and print
my_prediction = my_tree_one.predict(test_features)
print('This is the prediction:\n')
print(my_prediction)

# Create a data frame with two columns: PassengerId & Survived
# Survived contains the model's prediction
PassengerId =np.array(test["PassengerId"]).astype(int)
my_solution = pd.DataFrame(my_prediction, PassengerId, columns
= ["Survived"])
print('This is the solution in toto:\n')
print(my_solution)

# Check that the data frame has 418 entries
print('This is the solution shape:\n')
print(my_solution.shape)

# Write the solution to a CSV file with the name my_solution.csv
my_solution.to_csv("my_solution_one.csv," index_label =
["PassengerId"])
```

Here's the output:

```
These are the features:
[[   3.        1.        34.5       7.8292]
 [   3.        0.        47.        7.    ]
 [   2.        1.        62.        9.6875] ...,
 [   3.        1.        38.5       7.25  ]
 [   3.        1.        27.        8.05  ]
 [   3.        1.        27.        22.3583]]
```

This is the prediction:
[0 0 1 1 1 0 0 0 1 0 0 0 1 1 1 1 0 1 1 0 0 1 1 0 1 0 1 1 1 0 0 0 1 0 1 0 0
 0 0 1 0 1 0 1 1 0 0 0 1 1 1 0 1 1 1 0 0 0 1 1 0 0 0 1 0 0 1 0 0 1 1 0 0 0
 1 0 0 1 0 1 1 0 0 0 0 0 1 1 1 1 1 1 1 0 0 0 1 1 1 0 1 0 0 0 1 0 0 0 0 0 0
 0 1 1 1 0 1 1 0 1 1 0 1 0 0 1 0 1 0 0 1 0 0 1 0 0 1 0 0 0 0 0 0 0 0 1 1 0
 1 0 1 0 0 1 0 0 1 1 0 1 1 1 1 0 1 1 0 0 0 0 1 0 1 0 1 1 0 1 1 0 0 1 0 1
 0 1 0 0 0 0 0 1 0 1 0 1 0 0 0 0 1 0 1 0 0 0 0 1 0 1 1 0 1 0 0 1 0 1 0 1 0
 1 1 1 0 0 1 0 0 0 1 0 0 1 0 0 1 1 1 1 1 1 0 0 0 1 0 1 0 1 0 0 0 0 0 0 0 1
 0 0 0 1 1 0 0 0 0 0 0 0 0 1 0 1 1 0 0 0 0 0 1 1 0 1 0 0 0 1 0 1 0 1 0 0 0
 1 0 0 0 0 0 0 0 1 1 0 1 1 0 0 1 0 0 1 1 0 0 0 0 0 0 0 1 1 0 1 0 0 0 1 0 1
 1 0 0 0 0 0 1 0 0 0 1 0 1 0 0 0 1 1 0 0 0 1 0 1 0 0 1 0 1 1 1 1 0 0 0 1 0
 0 1 0 0 1 1 0 0 0 1 0 0 0 1 0 1 0 0 0 0 0 1 1 0 0 1 0 1 0 0 1 0 1 0 0 0 0
 0 1 1 1 1 0 0 1 0 0 0]
This is the solution in toto:

	Survived
892	0
893	0
894	1
895	1
896	1
897	0
898	0
899	0
900	1
901	0
902	0
903	0
904	1
905	1
906	1
907	1
908	0
909	1
910	1
911	0
912	0
913	1
914	1
915	0
916	1
917	0
918	1
919	1

```
920          1
921          0
...          ...
1280         0
1281         0
1282         0
1283         1
1284         1
1285         0
1286         0
1287         1
1288         0
1289         1
1290         0
1291         0
1292         1
1293         0
1294         1
1295         0
1296         0
1297         0
1298         0
1299         0
1300         1
1301         1
1302         1
1303         1
1304         0
1305         0
1306         1
1307         0
1308         0
1309         0
[418 rows x 1 columns]
This is the solution shape:
(418, 1)
```

That new column, Survived, contains a prediction for each of the 418 passengers listed in the test data set. We can write the predictions to a CSV file called *my_solution_one.csv*, upload the file to DataCamp, and verify that our predictions were 97 percent accurate. Ta-da! We just did machine learning. It was entry level, but it was machine learning nonetheless. When someone says they have "used artificial intelligence to make a decision," usually they mean "used machine learning," and usually they went through a process similar to the one we just worked through.

We created the Survived column and got a number that we can call 97 percent accurate. We learned that fare is the most influential factor in a mathematical analysis of *Titanic* survivor data. This was narrow artificial intelligence. It was not anything to be scared of, nor was it leading us toward a global takeover by superintelligent computers. "These are just statistical models, the same as those that Google uses to play board games or that your phone uses to make predictions about what word you're saying in order to transcribe your messages," Carnegie Mellon professor and machine learning researcher Zachary Lipton told the *Register* about AI. "They are no more sentient than a bowl of noodles."[17]

For a programmer, writing an algorithm is that easy. It gets made, it gets deployed, it seems to work. Nobody follows up. You maybe try turning the dials differently the next time to see if the accuracy seems to go up any. You try to get the highest number you can. Then, you move on to the next thing.

Meanwhile, out in the world, these numbers have consequences. It would be unwise to conclude from this data that people who pay more have a greater chance of surviving a maritime disaster. Nevertheless, a corporate executive could easily argue that it would be statistically legitimate to conclude this. If we were calculating insurance rates, we could say that people who pay higher ticket prices are less likely to die in iceberg accidents and thus represent a lower risk of early payout. People who pay more for tickets are wealthier than people who don't. This would allow us to charge rich people less for insurance. That's bad! The point of insurance is that risk is distributed evenly across a large pool of people. We've made more money for the insurance company, but we've not promoted the greatest good.

These types of computational techniques are used for *price optimization*, or grouping customers into very small segments to offer different prices to different groups. Price optimization is used in industries from insurance to travel—and it often results in price discrimination. A 2017 analysis by ProPublica and *Consumer Reports* found that in California, Illinois, Texas, and Missouri, some major insurers charged people who lived in minority neighborhoods as much as 30 percent more than people who lived in other areas with similar accident costs.[18] A 2014 analysis by the *Wall Street Journal* found that customers were being charged different prices for the same ordinary stapler on Staples.com. The price was higher or lower based on the customer's estimated zip code.[19] Christo Wilson, David Lazer, and a team of other Northeastern University researchers found different prices were offered to customers on Homedepot.com and on travel sites depending on whether the users viewed the sites on mobile devices or desktops.[20] Amazon admitted to experimenting with differential pricing in 2000. CEO Jeff Bezos apologized, calling it "a mistake."[21]

In an unequal world, if we make pricing algorithms based on what the world looks like, women and poor and minority customers inevitably get charged more. Math people are often surprised by this; women and poor and minority people are not surprised by this. Race, gender, and class influence pricing in a variety of obvious and devious ways. Women are charged more than men for haircuts, dry cleaning, razors, and even deodorant. Asian-Americans are twice as likely to be charged more for SAT prep courses.[22] African American restaurant servers make less in tips than white colleagues.[23] Being poor often means paying more for necessities. Furniture on installment plans costs more than outright purchase. Payday loans have a far higher interest rate than bank loans. Housing is considered affordable if it takes 30 percent or less of a household's monthly income, but poor renters are often stuck paying more for housing because of a variety of factors related to economic instability. "In Milwaukee, the majority of poor renters devote at least half their income to rent, and a third pay at least 80 percent," sociologist Pat Sharkey writes in a review of two ethnographies, Matthew Desmond's *Evicted: Poverty and Profit in the American City* and Mitchell Dunier's *Ghetto: The Invention of a Place, the History of an Idea.*[24] Inequality is unfair, but it's not uncommon. If machine-learning models simply replicate the world as it is now, we won't move toward a more just society. "The allure of the technology is clear—the ancient aspiration to predict the future, tempered with a modern twist of statistical sobriety," law professor and AI ethics expert Frank Pasquale writes in *The Black Box Society.* "Yet in a climate of secrecy, bad information is as likely to endure as good, and to result in unfair and even disastrous predictions."[25]

Part of the reason we run into problems when making social decisions with machine learning is that the numbers camouflage important social context. In the *Titanic* example, we picked a classifier, survival. We used features to predict our classifier, but there are other possible factors. For example, our *Titanic* dataset includes only age, sex, and the other factors.

We built our predictor based on the information we had. However, because this was a human and not a mathematical event, there were other factors at work.

Let's look at the night of the *Titanic* disaster. The *Titanic* received multiple warnings of ice from nearby ships over the course of the day on April 14, 1912. At 11:40 p.m., the ship hit an iceberg. Just after midnight, the *Titanic*'s captain, Edward John Smith, mustered the passengers and began to evacuate the ship. Smith issued an order: "Put the women and children in and lower away." First Officer William Murdoch was in charge of the lifeboats on the starboard side. Second Officer Charles Lightoller was in charge of the boats on the port side. Each man interpreted the captain's command differently. Murdoch thought the captain meant women and children *first*. Lightoller thought the captain meant women and children *only*. Murdoch let men onto the boats if all the nearby women and children had been loaded. Lightoller loaded all the women and children nearby, then lowered the boat even if it had empty seats. Both men let the boats down into the water even if the full capacity of sixty-five people had not been reached. There were not enough lifeboats for the people on board: *Titanic* carried only twenty boats for a ship rated to carry 3,547 people. The best records show that the ship carried a light load of 892 crew members and 1,320 passengers.

There is a potentially interesting test to be done on lifeboat numbers. Murdoch's boats on the starboard side had odd numbers; Lightoller's boats had even numbers. Men probably survived at a different rate according to their lifeboat number, because Lightoller, who was in charge of even-numbered boats, didn't load men. However, the lifeboat number isn't in the data. This is a profound and insurmountable problem. Unless a factor is loaded into the model and represented in a manner a computer can calculate, it won't count. Not everything that counts is counted. The computer can't reach out and find out the extra information that might matter. A human can.

There's also the problem of false causality. If we did have the lifeboat numbers, from a computational perspective it might look like men in odd-numbered lifeboats had a better chance of surviving the *Titanic* disaster. If we made decisions based on data, we might decide that all lifeboats should be odd-numbered so that we could save more men in case of emergency. Of course, this is ridiculous; it was the officer, not the number of the boat, that made the difference.

Two young men also confound the pure mathematical explanation. Walter Lord's *A Night to Remember*, a bestselling nonfiction account of the *Titanic* disaster, is a moving account of the ship's last hours.[26] Lord tells the story of Jack Thayer, a seventeen-year-old who boarded the *Titanic* in Cherbourg, France, after a long European holiday with his parents. Thayer made a friend on the ship, Milton Long, another young man his age traveling in first class. As the crisis on the ship intensified, both young men helped to get other passengers to safety. By 2:00 a.m., almost all the lifeboats had launched, with Long and Thayer handing women and children into the lifeboats. By 2:15 a.m., the last lifeboats had washed away in the swells. The ship was listing to port. There was an explosion; a wave crashed over the boat deck. Chef John Collins was standing on the boat deck holding a baby, helping a steward and a woman from steerage who was traveling with two children. He and the others were swept out to sea. The baby was torn out of his arms by the force of the wave.

Thayer and Long saw the chaos on the decks. Suddenly, the lights winked out; the water had reached the fireboxes in boiler room two. The only light came from the moon and the stars and the lanterns on the lifeboats slowly rowing away from the sinking ship. The second funnel collapsed with a crash. Thayer and Long looked around: the lifeboats were gone and no rescue ship was in sight. They realized the moment had come to jump. They shook hands. They wished each other good luck. Lord writes:

Long put his legs over the rail, while Thayer straddled it and began unbuttoning his overcoat. Long, hanging over the side and holding the rail with his hands, looked up at Thayer and asked, "You're coming, boy?"

"Go ahead, I'll be right with you," Thayer reassured him.

Long slid down, facing the ship. Ten seconds later Thayer swung his other leg over the rail and sat facing out. He was about ten feet above the water. Then with a push he jumped as far out as he could.

Of these two techniques for abandoning ship, Thayer's was the only one that worked.

Thayer survived by swimming to a nearby overturned lifeboat and clinging to it with forty others. He watched as the Titanic cracked in half, the bow and stern slipping under the water amid a field of debris. Thayer heard people crying in the water. It sounded like locusts, he thought. Eventually, lifeboat twelve picked up Thayer and the others from the icy water. Help arrived hours later. Thayer shivered in the lifeboat until 8:30 the next morning, when the passengers were rescued by the *Carpathia*.

Thayer and Long were young men of the same age, same physical ability, same social status, and absolutely the same opportunity to survive the disaster. The difference came down to a jump. Thayer leaped out as far as he could away from the ship; Long dropped down next to the ship. Long was sucked into the abyss; Thayer wasn't. What I find unsettling is that whatever the computer predicts for Thayer or Long, it will be wrong. The prediction is based only on fare class, age, and sex—but what really happened was a difference of jumps. The computer just fundamentally misunderstands. Long's death, the randomness of it, is why our statistical prediction of who survived and who died on the Titanic will never be 100 percent accurate—no statistical prediction can or will ever be 100 percent accurate—because human beings are not and never will be statistics.

This speaks to a principle called the *unreasonable effectiveness of data*. Unless you are alert to the possibilities of discrimination and disarray, AI seems like it works beautifully. One of my favorite explanations of the search to explain the world through computer science comes from a paper by Google researchers Alon Halevy, Peter Norvig, and Fernando Pereira. They write:

Eugene Wigner's article "The Unreasonable Effectiveness of Mathematics in the Natural Sciences" examines why so much of physics can be neatly explained with simple mathematical formulas such as $f=ma$ or $e=mc^2$. Meanwhile, sciences that involve human beings rather than elementary particles have proven more resistant to elegant mathematics. Economists suffer from physics envy over their inability to neatly model human behavior. An informal, incomplete grammar of the English language runs over 1,700 pages. Perhaps when it comes to natural language processing and related fields, we're doomed to complex theories that will never have the elegance of physics equations. But if that's so, we should stop acting as if or goal is to author extremely elegant theories, and instead embrace complexity and make use of the best ally we have: the unreasonable effectiveness of data.[27]

Data is unreasonably effective—seductively so, even. This explains why we can build a classifier that seems to predict with 97 percent accuracy whether a passenger survives the *Titanic* disaster and why a computer can defeat a human Go champion. It also explains why, when we look closely at what happens during the machine-learning process, the machine doesn't take into account any of the flukes that humans know happen in real-life disaster situations. Data is very effective. However, the data-driven approach ignores a number of factors that humans think matter a great deal.

Law and society are set up to accommodate all of the things that humans think matter. Data-driven decisions rarely fit with these complex sets of rules. The same unreasonable effectiveness of data appears in translation, voice-controlled smart home gadgets, and handwriting recognition. Words and word combinations are not understood by machines the way that humans understand them. Instead, statistical methods for speech recognition and machine translation rely on vast databases full of short word sequences, or *n-grams*, and probabilities. Google has been working on these problems for decades and has the best scientific minds on these topics, and they have more data than anyone has ever before assembled. The Google Books corpus, the *New York Times* corpus, the corpus of everything everyone has ever searched for using Google: it turns out that when you load all of this in and assemble a massive database of how often words occur near each other, it's unreasonably effective. Let's take something simple. In n-grams, the word *boat* usually occurs near *water*, so the two are probably related. The probability is higher that *boat* is closer to *water* than to *electorate* or *stink bug*, so a search pulls up terms or documents related to boats and water rather than to boats and stink bugs. People generally talk about the same types of things and search for the same types of things, and common knowledge is really quite common. The machine is not really learning; the search process is just inspired by human learning. If you read the math, which is all posted online, it's very clear that these calculations are not magic and are just math. The computer will get enough things right enough of the time that we may be tempted to call it mostly correct—but it will get things right for exactly the wrong reasons.

Because social decisions are about more than just calculations, problems will always ensue if we use data alone to make decisions that involve social and value judgments. Traveling first class on the *Titanic* meant someone was more likely to survive—but it would be wrong to deploy a model that suggests first-class travelers *deserve* to survive disasters more than people who travel second or third class. Nor should we do other things that derive from a flawed model like the one we created. Our *Titanic* model could be used to justify charging first-class passengers less for travel insurance, but that's absurd: we shouldn't penalize people for not being rich enough to travel first class. Most of all, we should know by now that there are some things machines will never learn and that human judgment, reinforcement, and interpretation is always necessary.

8   This Car Won't Drive Itself

The unreasonably effective data-driven approach works well enough for electronic search, for simple translation, and for simple navigation. Given enough training data, algorithms indeed will do a good job at a variety of mundane tasks, and human ingenuity usually fills in the blanks. With search, most of us by now have learned how to use increasingly complex or specific search terms (or at least synonyms) to find the specific web pages we're looking for when using a search box. Machine translation between languages is better than ever. It's still not as good as human translation, but human brains are magnificent at figuring out the meaning of garbled sentences. A stilted, awkward translation of a web page is usually all the casual web surfer needs. GPS systems that provide directions from point A to point B are terribly handy. They don't always give the *best* directions to the airport, if you ask any professional taxi or ride-share driver, but they will get you there and they will mostly show the traffic on the route for sufficiently busy areas.

However, the unreasonably effective data-driven approach has enough problems that I'm skeptical about using AI to fully replace humans for actual, life-threatening situations, like driving. The best case for considering how artificial intelligence works both really well and not at all is the case of the self-driving car.

The first time I rode in a self-driving car, in 2007, I thought I was going to die. Or vomit. Or both. So, when I heard in 2016 that self-driving cars were coming to market, that Tesla had created software called *Autopilot* and that Uber was testing self-driving cars in Pittsburgh, I wondered: What had changed? Did the reckless engineers I met in 2007 actually manage to embed an ethical decision-making entity inside a two-ton killing machine?

It turned out that perhaps not as much has changed as I might have thought. The story of the race to build a self-driving car is a story about the fundamental limits of computing. Looking at what worked—and what didn't—during the first decade of autonomous vehicles is a cautionary tale about how technochauvinism can lead to magical thinking about technology and can create a public health hazard.

My first ride happened on an autonomous vehicle test track: the weekend-empty parking lot of the Boeing factory in South Philadelphia. The Ben Franklin Racing Team, a group of engineering students at the University of Pennsylvania, was building an autonomous vehicle for a competition. I was writing a story about them for the University of Pennsylvania alumni magazine. I met the members of the Ben Franklin Racing Team on campus at dawn on a Sunday morning, and I followed them down the highway for self-driving-car racing practice.

The team had to practice at times when there was little traffic and few people around. Their car, a tricked-out Toyota Prius, wasn't exactly street legal. There are rules about what has to be in a car: a steering wheel, for example. They were OK to practice in the parking lot, or on Penn property, but the drive down I-95 from a West Philadelphia garage to the practice space in South Philly was a risk. They were less likely to get pulled over early on a Sunday morning because fewer police cars were patrolling the highway. The university lawyers were working on the state level to change legislation so that the car could legally drive itself. Until then, the racing team practiced at odd hours and hoped for the best.

I pulled up behind the Prius, which they had christened Little Ben, in the parking lot. The car was packed with engineers: mechanical-engineering student Tully Foote at the wheel, with electrical- and systems-engineering PhD candidate Paul Vernaza in the backseat next to Alex Stewart, a doctoral candidate in electrical engineering. Heteen Choxi, a Lockheed Martin employee and recent Drexel computer science grad, rode shotgun, wearing a bright yellow and black team jacket. As the car rolled to a gentle stop, Foote got out and popped the hatchback to reveal a mess of wires snaking over the back seat and onto the roof. The car looked like something out of a postapocalyptic movie, with sensors and miscellaneous parts bolted to the roof. The students had ripped a hole in the plastic console covering the dashboard. A tangle of wires spilled out, connected to a large, serious-looking laptop. Half of the trunk's floor was covered by Plexiglas, and more wires and boxes were visible in the wheel well. Foote pulled up a command prompt on an LCD screen installed in the trunk, and soon a satellite image of the parking lot popped into view. The three passengers remained belted in the car, each hunched over a laptop. Driving practice began.

In the competition, the 2007 Grand Challenge, Little Ben would need to drive itself through an empty "city" made out of a decommissioned military base. No remote controls, no preprogrammed paths through the city: just eighty-nine autonomous vehicles trying to drive down streets, around corners, through intersections, and around each other. The sponsor, the Defense Advanced Research Projects Agency (DARPA), promised a $2 million prize to the fastest finisher, plus $1 million and $500,000 prizes to the runners-up.

Robot-car technology was already assisting everyday drivers in 2007. By then, Lexus had released a car that could parallel-park itself under specific conditions. "Today, all of the high-end cars have features like adaptive cruise control or parking assistance. It's getting more and more automated," explained Dan Lee, associate professor of engineering and the team's adviser. "Now, to do it fully, the car has to have a complete awareness of the surrounding world. These are the hard problems of robotics: computer vision, having computers 'hear' sounds, having computers understand what's happening in the world around them. This is a good environment to test these things."

For Little Ben to "see" an obstacle and drive around it, the automated driving and GPS navigation had to work properly, and the laser sensors on the roof rack had to observe the object. Then, Little Ben had to identify the object as an obstacle and develop a path around it. One of the goals for that day's practice was to work on the subroutines that would eventually allow Little Ben to steer clear of other cars.

"The system is complicated enough that there are a lot of unforeseen consequences," said Foote. "If one thing runs slow, something else crashes. In software development in general, the standard is that you spend three-fourths of your time debugging. In a project like this, it's more like nine-tenths of the time debugging."

The 2007 challenge was more complicated than its predecessor. In the 2005 challenge, the task was to create a robot that could navigate 175 miles through the desert without human intervention in less than ten hours. On October 9, 2005, the Stanford Racing Team and their car Stanley won this competition (and the $2 million prize) for sending Stanley across 132 miles of the Mojave Desert. Stanley averaged 19 mph on the course, and finished in six hours, fifty-four minutes. In a desert, "it didn't really matter whether an obstacle was a rock or a bush because either way you'd drive around it," said Sebastian Thrun, then a Stanford associate professor of computer science and electrical engineering.[1] In the urban challenge, however, cars had to negotiate right-of-way and obey conventional rules of the road. "The challenge is to move from just sensing the environment to understanding the environment," said Thrun. Stanford's new Challenge car, Junior, a 2006 Volkswagen Passat, was considered a major rival to Little Ben. So was Boss, a 2007 Chevy Tahoe being developed by the Carnegie Mellon University (CMU) team. Carnegie Mellon fielded two cars in 2005, Sandstorm and H1ghlander, which placed second and third, respectively. The robotics rivalry between CMU and Stanford was akin to the basketball rivalry between University of North Carolina and Duke. Stanford poached Thrun, formerly CMU's star robotics professor, in 2003.

Back in the Boeing lot, senior electrical-engineering major Alex Kushleyev pulled up in his own brand-new car, a Nissan Altima. He had gone out to buy a remote control of the type used for toy cars: this was the emergency stop button. Every robot seems to include a large, cartoonish, red button. Two additional buttons were duct-taped to the rear side panels of the car and connected to the server rack of Mac Minis that made up the car's electronic "brain." The team had spent about $100,000 on the project to that point, through Penn's General Robotics, Automation, Sensing and Perception (GRASP) Laboratory. Lockheed Martin Advanced Technology Laboratories in Cherry Hill, New Jersey, and Maryland-based Thales Communications also sponsored the team.

"The Prius gives us more maneuverability, and since it is a hybrid car, it has a big on-board battery. We run a lot of computers, a lot of sensors, a lot of motors in addition to the car, so we need that extra power," said Lee. An electric motor controlled Little Ben's gas, brakes, and steering; all functions, from turn signals to wipers, could be controlled by buttons on a panel mounted above the gearshift, not unlike the customization used by disabled drivers who use their hands instead of their legs to drive. The car could be driven in the ordinary manner, or it could be driven using the hand controls. When the autopilot was engaged, they claimed there would be no need for a driver at all.

I watched the car travel short spurts through the parking lot. A safety driver sat in the passenger's seat with one hand on the emergency stop button. It was unsettling, but thrilling, to see the car driving ahead, with its steering wheel moving in front of an empty driver's seat.

As the battery made a low hum, Kushleyev took the wheel and drove across the parking lot at 15 mph. The day's goal was to rehearse Little Ben around parking lot obstacles. In the competition, Little Ben would have to navigate intersections and curbs and make decisions about how to react to stop signs, other cars, and stray dogs at a maximum speed of 30 mph.

Finally, it was my turn to take the wheel. I sat in the driver's seat. It felt curiously empty. Kushleyev turned on the automated-driving mechanism, and the car advanced a few feet—then lurched wildly to the left, then to the right, and jumped off its trajectory. "Gain control!" Stewart shouted from the backseat. The car headed toward a streetlight. As we neared the cement wall at the base of the light, the car accelerated. We were on a collision course. I jammed my foot down on what should have been the brake, only to find that it had been modified in a way I didn't understand. "Shouldn't this thing slow down?" I called out in a panic. I closed my eyes, certain that I was about to crash, and prepared to scream.

I heard murmurs and furious typing from the backseat. Kushleyev hit override, and the brakes. The car jerked to a stop, mere inches from the cement wall. My stomach felt like it had been left four feet behind.

I turned around to glare at the guy with the laptop. "There must be a bug in the program," he said with a shrug. "It happens."

"Only a GPS reading away from death," Stewart announced cheerfully. The engineers debated the swerving problem: the car was making a big wiggle where it was supposed to make a small, smooth turn. The laser sensors were scanning the area ahead of the car, but the software wasn't registering the light pole as an obstacle. This seemed to be affecting the steering, causing the car to jerk instead of turning smoothly.

Foote and Stewart conferred. They were robot-car veterans, having worked on two Grand Challenge robot cars as undergrads at Caltech. Their last autonomous vehicle was Alice, a Ford E350 van developed for the 2005 Grand Challenge. In the desert race, Alice drove herself about seven miles before heading into—and over—a barrier separating the media tent from the race course. Judges disabled Alice before she made headlines.

Little Ben's steering wheel moved on its own a few times; Stewart and Vernaza were controlling it from the backseat. Code problem solved, Kushleyev drove the car across the parking lot again and engaged the autopilot. The steering jerked, and the car headed toward an enormous snowplow parked at the edge of the lot as a grating sound screamed out of the engine.

"Bugger," said Stewart.

"Maybe it's Sheep?" said Vernaza, naming one of the programs controlling the car.

"This is high up on my list of things I don't want to fix today," said Stewart.

What I thought (but didn't write) at the time was that the experience did not inspire confidence in the technology. Riding in their car felt dangerous, like being in a car driven by a drunk toddler. If these were the folks making self-driving-car technology, their recklessness with my life did not augur well for the future. I couldn't see trusting my own child to a machine built by these kids. I didn't like the idea of this car being on the road; it seemed like a public menace. I wrote the story and assumed that the tech would fizzle out or be absorbed into another project, fading into tech obscurity like RealPlayer video or Macromedia Director or Jaz drives. After I filed my story, I forgot about the Penn robot car.

Meanwhile, Little Ben still had a race to win. On the morning of the DARPA Grand Challenge, November 3, 2007, the vehicles lined up at the starting gate. Their goal was to traverse the streets of George Air Force Base, a decommissioned military base in Nevada. There were roads and signs and escort vehicles. It was a motley crew of jerry-rigged vehicles lined up at the starting line. The task was to navigate sixty miles through the base, obeying street signs and avoiding other cars.

Pole position had been determined by qualifying runs the week before. Carnegie Mellon's car, Boss, was the top seed, meaning it could go into the course first, followed at intervals by the other robot cars and by some chase cars driven by humans. At the starting line, the Boss team was ready to go— but Boss wasn't. Its GPS wasn't working. A flurry of activity ensued. Other cars entered the course as Carnegie Mellon team members swarmed the car. Eventually, the source of the problem was identified: radio-frequency interference from the jumbotron television monitor located next to the race start chute. The jumbotron was jamming the GPS signals. Someone turned off the television.

Boss hit the streets tenth, twenty minutes behind the Stanford car. It was not a high-speed endeavor: Boss averaged about 14 mph over the fifty-five-mile course. "Everything that I saw Boss do looked great," said Chris Urmson, the team's director of technology. "It was smooth. It was fast. It interacted with other traffic well. It did what it was supposed to do."

Boss came in first. The Stanford team came in second, with a time about twenty minutes behind Boss. Little Ben finished the race, but not in the money. Teams from Cornell and MIT finished too, but not within the six-hour time limit of the race. It was clear that Pittsburgh and Palo Alto were the dominant powers in robot-car technology.

The difference between the Penn team's approach and the Stanford/CMU approach was significant. Little Ben's approach was knowledge-based. The team was trying to construct a machine that would decide what to do on the road based on a knowledge base and a set of programmed "experiences." This knowledge-based approach was one of the two major strains of artificial intelligence thinking. The Ben Franklin Racing Team was going for the general AI solution. It didn't work well enough.

Little Ben was trying to "see" obstacles the way a human might. The lidar, a laser radar mounted on the roof, would identify objects. Then, the software "brain" would identify the object based on criteria like shape, color, and size. It would go through a decision tree to decide what to do: if it is a living thing like a person or a dog, slow down; if it is a living thing like a bird, it will probably move out of the way, so no need to slow down. This required Little Ben to have a massive amount of information about objects in the real world. For example, consider a traffic cone. When upright, a traffic cone is notable for its triangular shape with a square base. Traffic cones are usually between twelve inches and 3.5 feet tall. We can write a rule that goes something like this:

```
identify object:
    IF object.color = orange AND object.shape = triangular_
        with_square_base
    THEN object = traffic_cone;
    IF object.identifier = traffic_cone
    THEN intitiate_avoid_sequence
```

What if the traffic cone's knocked over? I live in Manhattan; I see traffic cones knocked over all the time. I've seen a street blocked off by traffic cones, and I've seen people get out of their cars and move the traffic cones aside so they can drive down the street anyway. I've seen traffic cones mashed flat in the middle of the street. So, the rule about traffic cones has to be modified. Let's try something else:

```
identify object:
    IF object.color = orange AND object.shape is like
        triangular_with_square_base.rotated_in_3D
    THEN object = traffic_cone;
    IF object.identifier = traffic_cone
    THEN intitiate_avoid_sequence
```

Here, we run into a difference between human thought and computation. A human brain can rotate an object in space. When I say "traffic cone," you can picture the cone in your head. If I say, "imagine the cone is knocked over on the ground," you can probably imagine this too and can mentally rotate the object. Engineers are particularly good at imagining spatial manipulations in their heads. One popular math aptitude test for children involves showing them a 3-D shape on a 2-D plane, then presenting other pictures and asking them to choose which one represents the object rotated.

The computer has no imagination, however. To have a rotated image of the object, it needs a 3-D rendering of the object—a vector map, at the very least. The programmer needs to program in the 3-D image. A computer also isn't good at guessing, the way a brain is. The object on the ground is either something in its list of known objects, or it isn't.

Little Ben did two things when I rode in it: it drove in a circle, and it failed to avoid an obstacle. After I got over being terrified, I thought through the reasons that Little Ben didn't avoid the obstacle. The obstacle was a pillar. Little Ben needed a rule like "if obstacle.exists_in_path and obstacle.type=stationary, obstacle.avoid." However, this rule doesn't work because not all stationary objects remain stationary. A person might appear stationary for a moment, then the person might move. Therefore, the rule might be "if obstacle.exists_in_path and obstacle.type = stationary, AND obstacle.is_not_person, avoid." That doesn't work either: now we need to define the difference between a person and a column, so we're back to an object-classification problem. If the column can be recognized as a column, we could write a rule for columns and a rule for people. However, we don't know it's a column unless there's vision or, at the very least, object recognition—which is why I almost died in a car that almost ran into a giant cement pillar.

The core problem is sentience. Because there was no way to program theory of mind, the car would never be able to respond to obstacles the way that a human might. A computer only "knows" what it's been told. Without *sentience*, the cognitive capacity to reason about the future, it can't make the split-second decisions necessary to identify a streetlight as an obstacle and take appropriate evasive measures.

This problem of sentience was the central challenge of AI from its inception. It was the problem that Minsky eventually declared to be one of the hardest ever attempted. This is perhaps why, over at Stanford and Carnegie Mellon, they didn't even take it on. Their cars took a radically different approach to solving the problem of getting a vehicle through an obstacle course. Their narrow AI approach was purely mathematical, and relied on the unreasonable effectiveness of data. It worked better than anyone expected. I like to think of it as the *Karel the Robot plan*.

In 1981, Stanford professor Richard Pattis introduced an educational programming language called *Karel the Robot*.[2] Karel was named after Karel Čapek, the writer who invented the word "robot." Pattis's Karel the robot was not a real robot; he was an arrow drawn on a grid inside a square on a piece of paper. Students were supposed to pretend that the arrow was a robot in order to learn basic programming concepts. The box had one or more exits. Karel could move in the grid like a pawn in a chess game. Helping Karel escape from the box was the task. This introductory programming exercise—which you worked through with a pencil and paper—was the first computational assignment in programming classes at MIT, Harvard, Stanford, and all the other tech powerhouses for years. The professor gave us the box with Karel inside it. There were various obstacles. Our job as students was to write commands to get Karel from his location to the exit, avoiding the obstacles. It was mildly fun, or at least more fun than calculus, which was the other class I was taking when I was introduced to Karel in my freshman year. Here's an example of a Karel exercise (figure 8.1). The instructions for this puzzle read: "Every morning Karel is awakened in bed when the newspaper—represented by a beeper—is thrown on the front porch of the house. Program Karel to retrieve the paper and bring it back to bed. The newspaper is always thrown to the same spot, and Karel's world, including his bed, is as pictured."[2] Karel is represented by the arrow, and

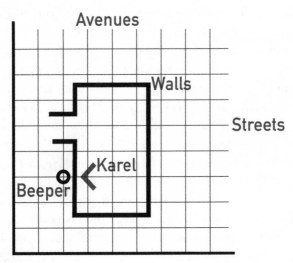

World
Borders

Figure 8.1
A typical Karel the Robot problem.

he is assumed to be in an imaginary bed in his initial position. To get to the beeper, he needs to turn ninety degrees north, travel north two streets, travel west two avenues, and so forth, until he reaches the beeper's address on the grid.

The key to solving Karel problems was knowing the obstacles in advance and routing Karel around them. The human programmer can see the grid, which is a map of Karel's entire world. Karel also has the grid stored in his internal memory; he is "aware" of the grid. The CMU team took a Karel approach when they built their car. They used the laser radar, cameras, and sensors on the car to build a 3-D map of the space. The map wasn't populated with "objects" that the car "recognized"; instead, it was populated with navigable and non-navigable areas that were identified using machine learning. Objects like other cars were rendered as 3-D blobs. The blobs were Karel-type obstacles.

This was brilliant because it cut down dramatically on the number of variables that Boss or Junior had to solve for. Little Ben had to identify all of the variables in sight—the road, the birds, the pedestrians, the buildings, the traffic cones—and then run a prediction for each variable's likely future location. It had to run a complex equation for each hypothetical. Boss and Junior didn't have to do this. Boss and Junior had been preloaded with a 3-D map of the landscape and the route they needed to navigate. Machine learning was used to identify in advance which parts of the 3-D map were navigable. The Junior/Boss approach was a narrow AI solution that relied on better mapping technology.

The car drove and created its own map of the environment. It made a grid, like Karel had a grid. Then, the car only had to consider the aberrations. If the traffic cone wasn't on the original map, it had to be factored in. If it was on the original grid, it was a stationary object and had been precalculated so that the processor didn't have to perform image recognition on the fly.

The CMU team had an edge over the other competitors. They had been working on computer-controlled vehicles for years already. ALVINN, a self-driving van, launched at CMU in 1989.[3] There was a stroke of enormous good fortune during the development period. Google founder Larry Page happened to become very interested in digital mapping. He attached a bunch of cameras to the outside of a panel van and drove around Mountain View, California, filming the landscape and turning the images into maps. Google then turned the van project into its massive Google Street View mapping program. Page's vision fit nicely with tech developed by the previously mentioned CMU professor Sebastian Thrun, who was active with the DARPA Challenge team. Thrun and his students developed a program that knit street photos together into maps. Thrun moved from CMU to Stanford. Google bought his tech and folded it into Google Street View.

Something important happened in hardware at this point too. Video and 3-D take up huge amounts of memory space. Moore's law says that the number of transistors on an integrated circuit doubles every year, and this increase in capacity means that computer memory has been getting cheaper and cheaper. Suddenly, around 2005, storage was cheap and abundant enough that for the first time it was feasible to make a 3-D map of the entire city of Mountain View and store it in a car's onboard memory. Cheap storage capacity was a game-changer.

Thrun and the other successful self-driving car engineers discovered that replicating the process of human perception and decision making is both devilishly complicated and impossible with current technology. They decided to ignore it. Usually, people invoke the Wright brothers at this point when talking about this kind of innovation. Before the Wright brothers, people thought that a flying machine had to mimic the action of a bird. The Wright brothers realized that they could make a flying machine without flapping—that gliding with wings was good enough.

The self-driving car programmers realized they could make a vehicle without sentience—that moving around in a grid is good enough. Their final design basically is a highly complicated remote-controlled car. It doesn't need to have awareness or to know rules for driving. What it uses instead are statistical estimates and the unreasonable effectiveness of data. It's an incredibly sophisticated cheat that's very cool and is effective in many situations, but a cheat nonetheless. It reminds me of using cheats to beat a video game. Instead of making a car that could move through the world like a person, these engineers turned the real world into a video game and navigated the car through it.

The statistical approach turns everything into numbers and estimates probabilities. Items in the real world are translated not into items, but into geometric shapes that move in certain directions on a grid at a calculated rate. The computer estimates the probability that a moving object will continue on its trajectory and predicts when the object will intersect with the vehicle. The car slows down or stops if the trajectories will intersect. It's an elegant solution. It gets approximately the correct result, but for the wrong reason.

This is a sharp contrast to how brains operate. From an *Atlantic* article in 2017: "Our brains today take in more than 11 million pieces of information at any given moment; because we can process only about 40 of those consciously, our nonconscious mind takes over, using biases and stereotypes and patterns to filter out the noise."[4]

How you feel about the car's autonomy depends on what you want to believe about AI. Lots of people, like Minsky and others, *want* to believe that computers can think. "We've had this A.I. fantasy for almost 60 years now," said Dennis Mortensen, x.ai's founder and CEO, to *Slate* in April 2016. "At every turn we thought the only outcome would be some human-level entity where we could converse with it like you and I are [conversing] right now. That's going to continue to be a fantasy. I can't see it in my lifetime or even my kids' lifetime."[5]

Mortensen said that what is possible is "extremely specialized, verticalized A.I.s that understand perhaps only one job, but do that job very well."

This is great—however, driving is not one job. Driving is many jobs simultaneously. The machine-learning approach is great for routine tasks inside a fixed universe of symbols. It's not great for operating a two-ton killing machine on streets that are teeming with gloriously unpredictable masses of people.

Since the 2007 Grand Challenge, DARPA has moved on from autonomous vehicles. Their current funding priorities don't include self-driving cars. "Life is by definition unpredictable. It is impossible for programmers to anticipate every problematic or surprising situation that might arise, which means existing ML systems remain susceptible to failures as they encounter the irregularities and unpredictability of real-world circumstances," said DARPA's Hava Siegelmann, program manager for the Lifelong Learning Machines Program, in 2017. "Today, if you want to extend an ML system's ability to perform in a new kind of situation, you have to take the system out of service and retrain it with additional data sets relevant to that new situation. This approach is just not scalable."[6]

However, the dream is alive in the commercial sphere. Today, decisions about autonomous vehicle rules are being left to the states. Nevada, California, and Pennsylvania are leading the pack, but at least nine other states have contemplated legislation that would permit some level of autonomous driving.

The fact that the decision is being left to the states is a huge problem. Having fifty different standards is practically impossible to program against. A programmer prefers to write once and run anywhere. If there are fifty states, plus Washington, DC, and the US territories, all of which have different traffic laws and standards for autonomous vehicles, programmers will have to rewrite traffic rules and operational rules for each state. We'll very quickly end up in the same confused, scattered situation that we face with missing textbooks in schools. States' rights are an important component of American democracy, but they are a monster to program against. Programmers don't even like to type; it's hard to imagine them being so detail-oriented that they voluntarily comply with fifty-plus different state traffic schemas and then manage to communicate these different operating procedures to each customer who buys an autonomous car.

The communication problem surfaces again when we talk about self-driving cars. The National Highway Traffic Safety Association (NHTSA), the government agency in charge of motor vehicle and highway safety, had to come up with a complex scale to describe autonomous driving so we could talk about it. For a long time, programmers and executives used the term *self-driving car* without defining specifically what they meant. Again—normal for language, problematic for policy. In an effort to wrangle the Wild West of autonomous vehicles, the NHTSA published a set of categories for autonomous vehicles. The September 2016 Federal Automated Vehicles Policy reads as follows:

There are multiple definitions for various levels of automation and for some time there has been need for standardization to aid clarity and consistency. Therefore, this Policy adopts the SAE International (SAE) definitions for levels of automation. The SAE definitions divide vehicles into levels based on "who does what, when."

Generally:

• At SAE Level 0, the human driver does everything;

• At SAE Level 1, an automated system on the vehicle can sometimes assist the human driver conduct some parts of the driving task;

• At SAE Level 2, an automated system on the vehicle can actually conduct some parts of the driving task, while the human continues to monitor the driving environment and performs the rest of the driving task;

• At SAE Level 3, an automated system can both actually conduct some parts of the driving task and monitor the driving environment in some instances, but the human driver must be ready to take back control when the automated system requests;

• At SAE Level 4, an automated system can conduct the driving task and monitor the driving environment, and the human need not take back control, but the automated system can operate only in certain environments and under certain conditions; and

• At SAE Level 5, the automated system can perform all driving tasks, under all conditions that a human driver could perform them.[7]

These standards changed at least once, and possibly twice, while I was writing this book—again, reminiscent of the changing school standards. At levels 3 and 4, the vehicle needs to sense its surroundings with complex, expensive sensors. The sensors used are primarily lidar, GPS, IMU, and cameras. The sensor input needs to be turned into binary information that is processed by the computer hardware inside the car. The hardware in this process is the same hardware that formed a "layer" of the turkey sandwich in chapter 2, and the same hardware that the Penn engineers wired into Little Ben's trunk. Each level requires increasing amounts of computing power to make driving decisions based on the input from the sensors. Nobody has managed yet to create hardware and software powerful enough to be safe for ordinary driving in all locations and weather conditions. "Currently, no vehicle commercially available today exceeds Level 2 autonomy," wrote Junko Yoshida in an October 2017 article about state-of-the-art computer chips for driving.[8] Level 5 doesn't exist for ordinary driving conditions and probably never will.

The great part of self-driving car development is the fact that driver-assistance technology has flourished. At levels 0–2, there have been a number of helpful innovations. People really like the idea of a car that can parallel-park itself—and as a small, finite exercise in geometry, it's a terrific use of technology.

Most of the autonomous vehicle research and some training data is available online in arXiv and in scholarly repositories.[9] On GitHub, there is training data available, and there is code that people are using for the Udacity open-source car competition (Thrun's latest venture). I looked at the Udacity image dataset. It had less information than I imagined it would. One major drawback to the data is that there's no weirdness built in, and the algorithms can't predict what isn't built in. Like in the *Titanic* data, there's no way to account for strategies of jumping off the sinking ship after all the lifeboats have departed.

In real life, weird stuff happens all the time. Former Waymo leader Chris Urmson, a Carnegie Mellon grad and Grand Challenge winner, laid out some of the strangest observations in a popular YouTube video. Waymo's test versions of automated cars have been driving around Mountain View for years, collecting data. Urmson laughed as he showed a bunch of kids playing Frogger across the highway or a woman in an electric wheelchair chasing a duck in circles around the middle of the road. These are not common occurrences, but they do happen. People have intelligence; they can accommodate weirdness. Computers aren't intelligent; they can't.

We can all think of strange things that we've observed while in the car. My own weirdest experience was with an animal. I was driving down winding mountain roads in Vermont with my friend Sarah, on our way to see a waterfall. We turned a blind corner, and there was a huge moose in the middle of the road. I skidded to a stop, my heart racing. I wondered how a self-driving car would handle this kind of situation. I went onto YouTube and watched some of the most popular fan videos of people playing with driver-assistance features. The videos I found were all made by men showing off their cool cars. They were uniformly positive. "It lulls you into a sense of security," a *Wired* writer said in a YouTube video about driving on a mostly empty highway in Nevada. He boasted about how little he had to do while using the Tesla Autopilot feature. Even though the directions clearly say to keep both hands on the wheel, he frequently bragged that he could take his hands off the wheel or just use one hand instead of two. He demonstrated some Easter eggs, jokes the programmers hid inside the code. He clicked six times on the steering wheel, and the display changed to show the rainbow road from Mario Kart. He showed a second Easter egg: the driver's display dinged for "more cowbell," a reference to a *Saturday Night Live* skit.

I watched some promotional videos for Waymo. In one, the narrator claimed Waymo's technology could "see" 360 degrees around the car, plus two football fields ahead. The shape of the car is optimized to allow field of view for the sensors. One major design feature, which isn't yet perfected, is that the computer must withstand vibrations and heat fluctuations. "We've been bolting things onto existing cars for a long time, and started to realize that's very limiting in what we can do when you're dealing with the constraint of an existing vehicle," said Jaime Waydo, a Waymo systems engineer, in a 2014 video. "When it comes to the physical operation of the vehicle, the sensors and the software are really doing all the work. There's no need for things like a steering wheel and a brake pedal, so all we really had to think about was a button to signal that we're ready to go. There's a lot of thought that goes into creating a prototype vehicle. We're learning a lot about safety."

Let's talk about safety. The major argument in favor of self-driving cars is that they will make the roads "safer." John Krafcik, the CEO of Waymo, has this on his LinkedIn page: "Globally, 1.2 million die each year in road accidents. 95% of the time, it's human error. Right now, there are about 1 billion cars on the planet. 95% of the time, they sit idly, wasting capital and consuming valuable space in our cities. We need to do better ... Self-driving cars could save thousands of lives, give people greater mobility, and free us from things we find frustrating about driving today."

Krafcik seems to be blaming drivers. Pesky humans, making human errors. This is technochauvinism. *Of course* humans are responsible for driving errors. Humans are the only ones driving cars! (Although I did once see what looked like a dog in a Yankees cap driving a miniature Mercedes down a sidewalk on Broadway in Lower Manhattan. I did a double take. The dog's owner was following behind with a remote control. This led to a delightful afternoon of exploring the subgenre of people who post online videos of animals in remote-controlled vehicles.)

We've had cars for a very long time, and we know that humans are going to make mistakes driving cars. They are human. Humans make mistakes. We know this. Even the humans who make software make mistakes. Nobody is a perfect driver. Even the people who write the software for autonomous vehicles are not perfect drivers. When you consider that humans drive trillions of miles every year, and avoid accidents for most of that time, it's quite impressive.

The same figure for human error comes up over and over. People dying is sad; I don't mean to minimize death. However, when you see a single statistic like this repeated, it raises suspicions. It usually means that it comes from a single source, meaning that it comes from a special interest group trying to influence public opinion. The figure that Krafcik quoted, 95 percent human error, also appears in a February 2015 report prepared by Santokh Singh, a senior mathematical statistician at Bowhead Systems Management, Inc., who was working under contract with the Mathematical Analysis Division of the National Center for Statistics and Analysis, an office of the NHTSA.[10] The report looks at a weighted sample of 5,470 crashes and assigns a cause to each one. The cause can be the driver, the car, or the environment (meaning the road or the weather).

Bowhead Systems Management is a subsidiary of Ukpeaġvik Iñupiat Corporation, a government contracting firm that manages the US Navy's unmanned autonomous system (UAS) operations in Maryland and Nevada. In other words: Bowhead, a company that makes unmanned autonomous systems for military use, has created the official government statistic that justifies building unmanned autonomous systems (cars) for civilian use.

The National Center for Health Statistics reports the number of deaths from motor vehicle traffic was 35,398 in 2014, the most recent year available. This is a rate of eleven deaths per one hundred thousand people. Overall, the age-adjusted death rate, which accounts for the aging of the population, was 724.6 deaths per 100,000 people.

Lots of people die in vehicle accidents; it's a major public health issue. In statistical lingo, dying from an injury is called *injury mortality*. Unintentional motor vehicle traffic–related injuries were the leading cause of injury mortality from 2002 to 2010, followed by unintentional poisoning. In 2015, the US Department of Transportation's NHTSA showed a 7.7 percent increase in motor vehicle traffic deaths in 2015. An estimated 35,200 people died from vehicle accidents in 2015, up from the 32,675 reported fatalities in 2014.

We could speculate on the causes. Texting and distracted driving are certainly contributing to the uptick in deaths. One straightforward solution would be to invest more in public transportation. In the Bay Area of California, public transportation is woefully underfunded. The last time I tried to take a subway at rush hour in San Francisco, I had to wait for three trains to pass before I could squeeze into a jam-packed car. On the roads, the situation is even worse. I'm not surprised that Bay Area programmers want to make self-driving cars so that they can do something other than sit in traffic. Based on my limited observations, commuting in the Bay Area means spending an awful lot of time sitting in traffic. However, public transportation funding is a complex issue that requires massive, collaborative effort over a period of years. It involves government bureaucracy. This is exactly the kind of project that tech people don't want to attack because it takes a really long time and it's complicated and there aren't easy fixes.

Meanwhile, the self-driving car remains a fantasy. In 2011, Sebastian Thrun launched Google X, the company's "moonshot" division. In 2012, he founded Udacity, which also failed. "I'd aspired to give people a profound education—to teach them something substantial. But the data was at odds with this idea," Thrun told *Fast Company*. "We have a lousy product."[11]

Thrun has been honest about things he's tried that don't work—but it seems nobody's listening. Why not? The simplest explanation may be greed. Tech investor Roger McNamee told the *New Yorker*: "Some of us actually, as naïve as it sounds, came here to make the world a better place. And we did not succeed. We made some things better, we made some things worse, and in the meantime the libertarians took over, and they do not give a damn about right or wrong. They are here to make money."[12]

Finally, in 2017, curious to see how reality measured up to what I was reading, I tried to make an appointment to ride in a self-driving car. I tried Uber first; Pittsburgh isn't too far away from where I live. The PR person said there weren't any appointments available. I asked if I could just go to Pittsburgh and hail one. The PR person discouraged me. I realized why: the cars aren't in wide use. They're not ready for prime time.

Self-driving cars have problems: They don't track the center line of the street well on ill-maintained roads. They don't operate in snow and other bad weather because they can't "see" in these conditions. The lidar guidance system in an autonomous car works by bouncing laser beams off nearby objects. It estimates how far the objects are by measuring the reflection time. In the rain or snow or dust, the beams bounce off the particles in the air instead of bouncing off obstacles like bicyclists. One self-driving car was spotted going the wrong way down a one-way street. The software apparently didn't reflect that the street was one way. The cars are easy to confuse because they rely on the same mediocre image recognition algorithms that mislabel pictures of black people as gorillas.[13] Most autonomous vehicles use algorithms called deep neural networks, which can be confused by simply putting a sticker or graffiti on a stop sign.[14] GPS hacking is a very real danger for autonomous vehicles as well. Pocket-size GPS jammers are illegal, but they are easy to order online for about $50. Commercial truckers commonly use jammers in order to pass for free through GPS-enabled toll booths.[15] Self-driving cars navigate by GPS; what happens when a self-driving school bus speeding down the highway loses its navigation system at 75 mph because of a jammer in the next lane?

In the scientific community, there is an undercurrent of skepticism. An AI researcher told me: "I have a Tesla. The Autopilot ... I only use it for highway driving. It doesn't work for city driving. The technology isn't there yet. At NVIDIA, they found that self-driving car algorithms mess up an average of every ten minutes." This observation is consistent with the Tesla user's manual, which states that the Autopilot should only be used for short periods of time on highways under driver supervision.

Uber received bad press in 2017 after its then-CEO, Travis Kalanick, was filmed yelling at Uber driver Fawzi Kamal. Kamal had lost $97,000 and said he was bankrupt because of Uber's business strategy of cutting fares so that drivers make as little as ten dollars per hour. At the time, Kalanick's net worth was $6.3 billion. Kamal told Kalanick about his struggle. Kalanick replied: "Some people don't like to take responsibility for their own shit. They blame everything in their life on somebody else. Good luck!" The company launched self-driving cars in California in defiance of state regulations. They were shut down after a legal fight. Kalanick personally hired Anthony Levandowski, a DARPA Grand Challenge participant who worked with Thrun at Google X and later Waymo. Levandowski was fired from Uber in May 2017 for failing to cooperate with an investigation into whether he stole intellectual property from Waymo and used it to further Uber's technology interests.[16]

In May 2016, Joshua D. Brown of Canton, Ohio, became the first person to die in a self-driving car. Brown, forty years old, was a Navy SEAL and master Explosive Ordinance Disposal (EOD) technician turned technology entrepreneur. He died in his Tesla while using the Autopilot self-driving feature. He had such faith in his car that he was letting the Autopilot do all of the work. It was a bright day, and the car's sensors failed to pick up a white semi-tractor-trailer that was turning through an intersection. The Tesla drove under the truck. The entire top of the car sheared off. The base of the car kept going, coming to rest several hundred yards away.[17]

"When used in conjunction with driver oversight, the data is unequivocal that Autopilot reduces driver workload and results in a statistically significant improvement in safety," Tesla said in a statement after the crash.[18] The Associated Press wrote of the crash:

This is not the first time automatic braking systems have malfunctioned, and several have been recalled to fix problems. In November, for instance, Toyota had to recall 31,000 full-sized Lexus and Toyota cars because the automatic braking system radar mistook steel joints or plates in the road for an object ahead and put on the brakes. Also last fall, Ford recalled 37,000 F-150 pickups because they braked with nothing in the way. The company said the radar could become confused when passing a large, reflective truck.

The technology relies on multiple cameras, radar, laser and computers to sense objects and determine if they are in the car's way, said Mike Harley, an analyst at Kelley Blue Book. Systems like Tesla's, which rely heavily on cameras, "aren't sophisticated enough to overcome blindness from bright or low contrast light," he said.

Harley called the death unfortunate, but said that more deaths can be expected as the autonomous technology is refined.[19]

The NHTSA investigated the crash and all but said the crash was Brown's fault, not the computer's. They did note, however, that Tesla might reconsider its decision to call the feature *Autopilot*.

Decisions about how autonomous vehicles should react are quite literally life-and-death decisions. The Tesla Model X P90D has a curb weight of 5,381 pounds. For reference, a female Asian elephant weighs about six thousand pounds.

After I failed to book a ride in an Uber self-driving car in Pittsburgh, I tried to schedule a ride at NVIDIA, the company that makes the chips used in autonomous vehicles. They told me that it was a bad time and to check back after the Consumer Electronics Show (CES), a big tradeshow in Las Vegas. I did so; they didn't get back to me. Waymo said on its website that it isn't accepting press requests. Finally, to see the state of the art from a consumer perspective, I booked a test drive in a Tesla. On a bright, sunny, crisp winter morning, my family went to the Tesla dealer in Manhattan. The showroom is in the Meatpacking District, under the High Line, on West Twenty-Fifth Street. It's surrounded by art galleries that used to be auto body shops. Across the street was a wrought-iron outline of a woman. Someone had yarn-bombed it, covering the metal. A peach-colored crochet bikini hung limply on the frame.

We walked in and saw a red Model S sedan. Next to it on the floor was a miniature version, in the identical deep crimson. It was a Radio Flyer version of a Tesla Model S. It was miniature like a Barbie Jeep or a mini John Deere tractor or a Power Wheels car—but it was a Tesla. I was enchanted.

We went out on a test drive in the Model X with a salesperson named Ryan. The X doors open like falcon wings: they furl and unfurl. My son walked up to the car. Ryan beeped the remote, which was shaped like a small Tesla, to open the rear passenger side door. The door opened slowly, halfway. "It senses you standing there," Ryan said. "It won't flip open fast and hit you because it sees you." The door stopped. It wouldn't open all the way. Ryan beeped the remote again, looked concerned. He went to investigate. We stood on the sidewalk, watching.

Ryan returned, looking relieved. "It's the sensor," Ryan explained. The door sensor was right next to a sign reading "No parking on street cleaning days." The green metal pole was right next to the sensor, which is in the housing over the rear passenger side wheel. The sensor is one of eight cameras mounted in the car body. Because of the pole, Ryan said, the door wouldn't open all the way. He promised that we could take a picture with the wings up when we got back, when the car would be parked differently.

We got in and Ryan explained where everything was. I inhaled deeply. It smelled like new car and luxury. White "vegan leather" wrapped around the driver's seat and the seat back was encased in a shiny black plastic shell that looked like something out of a 1960s James Bond film. Really, the whole experience felt very James Bond.

I put my foot on the brake and the car started. Where a normal car would have buttons, the Model X has a giant touchscreen. There are only two buttons. One is for the hazard lights: "That's federally mandated," Ryan said, waving his hand apologetically. The other button is on the right side of the giant touchscreen. It opens the glove compartment.

The ride in an electric car doesn't jolt you like a car with a gasoline engine. There is a subtle shaking that happens with a gas engine. In the Tesla, this vibration is gone. The car felt quiet and smooth as we pulled out of the parking spot and drove toward the West Side Highway.

I tried to turn on the Autopilot feature via a lever to the left of the steering wheel. I pulled it toward me twice to activate. It beeped, and an orange light on the console blinked. "This is a new car. Autopilot isn't active," Ryan explained. "A massive Autopilot update just rolled out a few days ago. It will be another few weeks before the Autopilot works on this car. It needs to gather data."

"So it doesn't work?" I asked.

"It works," Ryan said. "The car is ready for total autonomy, but we can't implement it yet because, you know, regulations." "You know, regulations," meant that Joshua Brown died in an Autopilot crash and the NTHSA hadn't yet finished its investigation—so Tesla had turned off the Autopilot on all cars until the developers could build, test, and roll out new features.

Ryan chatted about the future, which in his view meant Teslas everywhere. "When we have full autonomy, Elon Musk says you should be able to press a button and summon your car no matter where you are. It might take a few days for your car to find you, but it should arrive." I wondered if it occurred to him that waiting for days for your car to arrive kind of defeats the point of having a car at all.

"Someday" is the most common way to talk about autonomous vehicles. Not if, but when. This seems strange to me. The fact that I couldn't get a ride at Uber or NVIDIA or Waymo means the same thing as Ryan's, "you know, regulations": The self-driving car doesn't really work. Or, it works well enough in easy driving situations: a clear day, an empty highway with recently painted lines. That's how Uber's subsidiary Otto (which was founded by Levandowski) managed to pull off a publicity stunt of sending a self-driving beer truck from the East Coast to the West Coast. If you set up the conditions just right, it looks like it works. However, the technical drawbacks are abundant. Continuous autonomous driving requires two onboard servers—one for operation, one for backup—and together the servers generate about five thousand watts. That wattage generates a lot of heat. It's the wattage you'd require to heat a four-hundred-square-foot room. No one has yet figured out how to also incorporate the cooling required to counter this.[20]

Ryan directed me onto the West Side Highway and into traffic. Ordinarily, I take my foot off the accelerator and drift up to a stop light, braking before I get there—but the Tesla has regenerative braking, which meant that the brakes kicked in when I took my foot off the gas pedal. It felt disorienting, the need to drive differently. Someone honked at me. I couldn't tell if he honked because I was being weird about the traffic light, if he was giving me a hard time for being in a luxury car, or if he was just an ordinary NYC jerk.

I drove down the highway and turned onto cobblestone-paved Clarkson Street. It felt less bumpy than usual. Ryan directed me just past Houston onto a block-long stretch of smooth road without any driveways and with few pedestrians that stretches along the back of a shipping facility. "Open it up," Ryan urged me. "There's nobody around. Try it."

I didn't need to be told twice. I pressed the pedal to the metal—I had always wanted to do that—and the car surged ahead. The power was intoxicating. We were all thrust back against the seats with the force of acceleration. "Just like Space Mountain!" said Ryan. My son, in the back seat, agreed. I regretted that the block was so short. We turned back onto the West Side Highway and I hit the accelerator again, just to feel the surge. Everyone was thrown back against the seats again.

"Sorry," I said. "I love this."

Ryan nodded reassuringly. "You're a very good driver," he told me. I beamed. I realized he probably says this to everyone, but I didn't care. My husband, I noticed in the rearview mirror, looked a little green.

"This is the safest car on the market," Ryan said. "The safest car ever made." He told a story about the NHTSA's crash testing on the Tesla: it couldn't crash it. "They tried to flip it—and they couldn't. They had to get a forklift to flip it over. We did the crash test, where the car drives into a wall—we broke the wall. They dropped a weight on the car, we broke the weight. We've broken more pieces of test equipment than any car ever."

We passed another Tesla in Greenwich Village, and we waved. This is a thing that Tesla owners do: they wave to each other. Drive a Tesla on the highway in San Francisco, and your arm gets tired from waving.

Ryan kept referring to Elon Musk. A cult of personality surrounds Musk, unlike any other car designer. Who designed the Ford Explorer? I have no idea. But Elon Musk, even my son knew. "He's famous," my son said. "He was even a guest star on the *Simpsons*."

We parked and took a picture of my son and me standing next to the bright white car, its wings up. We got into our family car parked outside. "This feels so old-fashioned now," my son said. We drove home down the West Side Highway, then over the cobblestones of Clarkson Street. We jolted and bobbled over the stones. It was the exact opposite of the smooth ride we felt in the Tesla. My car felt like it was shaking me at a low level. It was like the time I went to Le Bernadin for lunch, then came home and realized the only thing we had for dinner was hot dogs.

As a car, the Tesla is amazing. As an autonomous vehicle, I am skeptical. Part of the problem is that the machine ethics haven't been finalized because they are very difficult to articulate. The ethical dilemma is generally led by the trolley problem, a philosophical exercise. Imagine you're driving a trolley that's hurtling down the tracks toward a crowd of people. You can divert it to a different track, but you will hit one person. Which do you choose: certain death for one, or for many? Philosophers have been hired by Google and Uber to work out the ethical issues and embed them in the software. It hasn't worked well. In October 2016, *Fast Company* reported that Mercedes programmed its cars to always save the driver and the car's occupants.[21] This is not ideal. Imagine an autonomous Mercedes is skidding toward a crowd of kids standing at a school bus stop next to a tree. The Mercedes's software will choose to hit the crowd of children instead of the tree because this is the strategy that is most likely to ensure the safety of the driver—whereas a person would likely steer into the tree, because young lives are precious.

Imagine the opposite scenario: the car is programmed to sacrifice the driver and the occupants at the expense of bystanders. Would you get into that car with your child? Would you let anyone in your family ride in it? Do you want to be on the road, or on the sidewalk, or on a bicycle, next to cars that have no drivers and have unreliable software that is designed to kill you or the driver? Do you trust the unknown programmers who are making these decisions on your behalf? In a self-driving car, death is a feature, not a bug.

The trolley problem is a classic teaching example of computer ethics. Many engineers respond to this dilemma in an unsatisfying way. "If you know you can save at least one person, at least save that one. Save the one in the car," said Christoph von Hugo, Mercedes's manager of driverless car safety, in an interview with *Car and Driver*.[22] Computer scientists and engineers, following the precedent set by Minsky and previous generations, don't tend to think through the precedent that they're establishing or the implications of small design decisions. They ought to, but they often don't. Engineers, software developers, and computer scientists have minimal ethical training. The Association for Computing Machinery (ACM), the most powerful professional association in computing, does have an ethical code. In 2016, it was revised for the first time since 1992. The web, remember, launched in 1991 and Facebook launched in 2004.

There's an ethics requirement in the recommended standard computer science curriculum, but it isn't enforced. Few universities have a course in computer or engineering ethics on the books. Ethics and morality are beyond the scope of our current discussion, but suffice it to say that this isn't new territory. Moral considerations and concepts like the social contract are what we use when we get to the outer limits of what we know to be true or what we know how to deal with based on precedent. We imagine our way into a decision that fits with the collective framework of the society in which we live. Those frameworks may be shaped by religious communities or by physical communities. When people don't have a framework or a sense of commitment to others, however, they tend to make decisions that seem aberrant. In the case of self-driving cars, there's no way to make sure that the decisions made by individual technologists in corporate office buildings will match with actual collective good. This leads us to ask, again: Who does this technology serve? How does it serve us to use it? If self-driving cars are programmed to save the driver over a group of kindergarteners, why? What does it mean to accept that programming default and get behind the wheel?

Plenty of people, including technologists, are sounding warnings about self-driving cars and how they attempt to tackle very hard problems that haven't yet been solved. Internet pioneer Jaron Lanier warned of the economic consequences in an interview:

The way self-driving cars work is big data. It's not some brilliant artificial brain that knows how to drive a car. It's that the streets are digitized in great detail. So where does the data come from? To a degree, from automated cameras. But no matter where it comes from, at the bottom of the chain there will be someone operating it. It's not really automated. Whoever that is—maybe somebody wearing Google Glass on their head that sees a new pothole, or somebody on their bike that sees it—only a few people will pick up that data. At that point, when the data becomes rarified, the value should go up. The updating of the input that is needed is more valuable, per bit, than we imagine it would be today.[23]

Lanier is describing a world in which vehicle safety could depend on monetized data—a dystopia in which the best data goes to the people who can afford to pay the most for it. He's warning of a likely future path for self-driving cars that is neither safe nor ethical nor toward the greater good. The problem seems to be that few people are listening. "Self-driving cars are nifty and coming soon" seems to be the accepted wisdom, and nobody seems to care that the technologists have been saying "coming soon" for decades now. To date, all self-driving car "experiments" have required a driver and an engineer to be onboard at all times. Only a technochauvinist would call this success and not failure.

A few useful consumer advances have come out of self-driving car projects. My car has cameras embedded in all four sides; the live video from these cameras makes it easier to park. Some luxury cars now have a parallel-parking feature to help the driver get into a tight space. Some cars have a lane-monitoring feature that sounds an alert when the driver strays too close to the lane markings. I know some anxious drivers who really value this feature.

Safety features rarely sell cars, however. New features, like onboard DVD players and in-car Wi-Fi and integrated Bluetooth, are far more helpful in increasing automakers' profits. This is not necessarily toward the greater good, however. Safety statistics show that more technology inside cars is not necessarily better for driving. The National Safety Council, a watchdog group, reports that 53 percent of drivers believe that if manufacturers put infotainment dashboards and hands-free technology inside cars, these features must be safe to use. In reality, the opposite is true. The more infotainment technology goes into cars, the more accidents there are. Distracted driving is up since people started texting on mobile phones while driving. More than three thousand people per year die on US roads in distracted driving accidents. The National Safety Council estimates that it takes an average of twenty-seven seconds for the driver's full mental attention to return after checking a phone. Texting while driving is banned in forty-six states, the District of Columbia, Puerto Rico, Guam, and the US Virgin Islands. Nevertheless, drivers persist in using phones to talk or text or find directions while driving. Young people are particularly at fault. Between 2006 and 2015, the number of drivers aged sixteen to twenty-four who were visibly manipulating handheld devices went up from 0.5 percent to 4.9 percent, according to the NHTSA.[24]

Building self-driving cars to solve safety problems is like deploying nano-bots to kill bugs on houseplants. We should really focus on making human-assistance systems instead of on making human-replacement systems. The point is not to make a world run by machines; people are the point. We need human-centered design. One example of human-centered design might be for car manufacturers to put into their standard onboard package a device that blocks the driver's cell phone. This technology already exists. It's customizable so that the driver can call 911 if need be but otherwise can't call or text or go online. This would cut down on distracted driving significantly. However, it would not lead to an economic payday. The hope of a big payout is behind a great deal of the hype behind self-driving cars. Few investors are willing to give up this hope.

The economics of self-driving cars may come down to public perception. In a 2016 conversation between President Barack Obama and MIT Media Lab director Joi Ito, which was published in *Wired*, the two men talked about the future of autonomous vehicles.[25] "The technology is essentially here," Obama said.

We have machines that can make a bunch of quick decisions that could drastically reduce traffic fatalities, drastically improve the efficiency of our transportation grid, and help solve things like carbon emissions that are causing the warming of the planet. But Joi made a very elegant point, which is, what are the values that we're going to embed in the cars? There are gonna be a bunch of choices that you have to make, the classic problem being: If the car is driving, you can swerve to avoid hitting a pedestrian, but then you might hit a wall and kill yourself. It's a moral decision, and who's setting up those rules?

Ito replied: "When we did the car trolley problem, we found that most people liked the idea that the driver and the passengers could be sacrificed to save many people. They also said they would never buy a self-driving car." It should surprise no one that members of the public are both more ethical and more intelligent than the machines we are being encouraged to entrust our lives to.

9   Popular Doesn't Mean Good

How can you take a "good" selfie? In 2015, several prominent American media outlets covered the results of an experiment that purported to answer this question using data science. The results were predictable to anyone familiar with photography basics: make sure your picture is in focus, don't cut off the subject's forehead, and so forth. The experiment used the same type of procedures we used to analyze the *Titanic* data in chapter 7.

What was notable about the experiment—but was not noted by the investigator, Andrej Karpathy, then a Stanford PhD student and now the head of AI at Tesla—was that almost all the "good" photos were of young white women, despite the fact that older women, men, and people of color were included in the original pool of selfies. Karpathy used a measure of *popularity*—the number of "likes" each photo garnered on social media—as the metric for what constituted *good*. This type of mistake is quite common among computational researchers who do not critically reflect on the social values and human behaviors that lead to statistics being produced. Karpathy assumed that the photos were popular, and therefore they must be good. By selecting for popularity, the data scientist created a model that had significant bias: it prioritized young, white, cisgender images of women that fit a narrow, heteronormative definition of attractiveness. Let's say that you are an older black man, and you give your selfie to Karpathy's model to be rated. The model will not label your photo as good, no matter what. You are not white and you are not a cisgender woman and you are not young; therefore you do not satisfy the model's criteria for "good." The social implication for a reader is that unless you look a certain way, your picture cannot possibly be good. This is not true. Also, no kind or reasonable person would say this to another person!

This conflation of *popular* and *good* has implications for all computational decision making that involves subjective judgments of quality. Namely: a human can perceive a difference between the concepts *popular* and *good*. A human can identify things that are popular but not good (like ramen burgers or racism) or good but not popular (like income taxes or speed limits) and rank them in a socially appropriate manner. (Of course, there are also things like exercise and babies that are both popular and good.) A machine, however, can only identify things that are popular using criteria specified in an algorithm. The machine cannot autonomously identify the quality of the popular items.

This brings us back to the fundamental problem: algorithms are designed by people, and people embed their unconscious biases in algorithms. It's rarely intentional—but this doesn't mean we should let data scientists off the hook. It means we should be critical about and vigilant for the things we know can go wrong. If we assume discrimination is the default, then we can design systems that work toward notions of equality.

One of the core values of the Internet is the idea that things can be ranked. Today's society is mad for measurement; it's unclear to me whether the mania for measurement arose from the mathematical frenzy for ranking, or if the mathematical frenzy is simply a response to a social incentive. In either case, ranking is king right now. We have college rankings, sports team rankings, and hackathon team rankings. Students jockey for class rank position. Schools are ranked. Employees are ranked.

Everybody wants to be at the top; nobody wants to be in the bottom, and nobody wants to hire (or select) from the bottom. However, in education, which is the area that I know best, there is a logical fallacy at work. If we look at a pool of one thousand students and their test scores, usually the scores will fit a bell curve. Half the students will be above average, half will be below, and there will be a small percentage who score really well or who score terribly. That's normal—but school districts and state officials insist that their goal is to have all students at a level of "competence." This is impossible unless you set the bar for competence at zero. It's quite popular for school districts to claim they want all of their students to be high-achieving, but it isn't necessarily good to strive toward an impossible ideal.

The idea that popular is more important than good is baked into the very DNA of Internet search. Consider the origin of search: Back in the 1990s, two computer science graduate students wondered what to read next. Their discipline was only fifty years old (as opposed to the hundreds of years of history of their brethren in mathematics). It was difficult to figure out what to read outside of the syllabi handed out in class.

They had read some math about analyzing citations to get a citation index, and they decided to try to apply this math to web pages. (There weren't very many web pages at this point.) Their problem was how to identify "good" web pages, meaning web pages that they thought would be worth reading. The idea was that it would be just like academic citations: In computer science, the most-often-cited papers are the most important. By definition, the good papers become the most popular. Therefore, they built a search engine that would calculate how many incoming links pointed to a given web page, then they ran an equation to generate a ranking called PageRank based on the number of incoming links and the ranking of the outgoing links on a page. They reasoned that web users would act just like academics: each web user would create web pages that linked to other pages that each user considered good. A popular page, one with a large number of incoming links, was ranked higher than a page with fewer incoming links. PageRank was named after one of the grad students, Larry Page. Page and his partner, Sergei Brin, went on to commercialize their algorithm and create Google, one of the most influential companies in the world.

For a long time, PageRank worked beautifully. The popular web pages were the good ones—in part because there was so little content on the web that *good* was not a very high threshold. However, more and more people went online, content swelled, and Google began to make money based on selling advertising on web pages. The search-ranking model was taken from academic publishing; the advertising model was taken from print publishing.

As people learned how to game the PageRank algorithm to elevate their position in search results, popularity became a kind of currency on the web. Google engineers had to add factors to search so that spammers wouldn't game the system. They added multiple features, iterating and tweaking the algorithm. One interesting feature is the geolocation dimension they added to autocomplete what the user types in the search box. Search autocomplete is based on what's happening around you. If you type "ga" into the search box, it will autocomplete with "GA" if lots of people near you are searching for Georgia topics (or possibly UGA football) or with "Lady Gaga" if lots of people around you are searching for the musician. Now, there are over two hundred factors that go into search, and PageRank has been augmented by many additional methods, including machine learning. It works beautifully—except when it doesn't.

A good example of how tech doesn't quite translate comes from how page designers create the front page of a newspaper. It's highly curated. The different areas have names: *above the fold* and *below the fold* are the most obvious. In the *Wall Street Journal* (WSJ), there's always a bright spot on the front page, called the *A-hed*. It brings levity. Longtime WSJ staffer Barry Newman writes:

The "A-hed" started out as just another headline code. It soon became the code name for a story light enough to "float off the page." The A-hed is a headline that doesn't scream. It giggles.

Great editors, it's been said, create vessels into which writers can pour their work. That's what Barney Kilgore did starting in 1941. The modern *Journal*'s first managing editor, he knew that into the world of business a little mirth must be poured.

By putting the fun out front, wrapped around the day's woes, Kilgore sent a larger message: That anyone serious enough about life to read the *Wall Street Journal* should also be wise enough to step back and consider life's absurdities ...

Done well, an A-hed is more than a news feature. Ideas rise out of our personalities, our curiosities and our passions. A-heds aren't humor columns. They don't push opinions. We don't make stuff up. Sometimes, a touch of poignancy can set all joking aside. Yet two reporters reporting on the same oddity will always report its oddness in their own odd ways.[1]

This is substantially different than a scroll like the Facebook news feed because a page editor will consider a mix: something light, something dark, and a few mid-range stories to balance. A front page is a precise mix of elements. The *New York Times* has a team that curates its digital front page manually all day, every day. Few other news organizations can afford this staff effort, so at smaller outlets a homepage might be curated once a day or automatically populated based on the print front page. The page editor's curation adds value to the reading experience. This is good, but not popular: news homepage traffic has been steadily declining since social media began to eat the world.

It's popular to blame journalists, and journalism, for the decline of public discourse. I would argue that such blame is misplaced and not good for society, however. The switch from print to digital has had a dramatic impact on the quality of journalism produced in the United States. The Bureau of Labor Statistics reports that among information industries in

2015, average annual pay in the Internet publishing and web search portals industry was $197,549. Average annual pay was only $48,403 for those in newspaper publishing and $56,332 for radio broadcasting.[2] As newsrooms empty out because talented writers and investigative journalists are seeking higher-paying jobs, there are fewer people left to keep the foxes out of the henhouse.

This is a problem because cheating is baked into the DNA of modern computer technology and modern tech culture. Around 2002, when Illinois redesigned the image that would be imprinted on its quarter-dollar coins as part of the nationwide quarter redesign, state officials decided to hold a contest so that citizens could vote for the design they liked best. A programmer friend of mine had a clear favorite: the Land of Lincoln design, which showed a handsome young Abraham Lincoln holding a book inside an outline of the state of Illinois. To Lincoln's left was a silhouette of the Chicago skyline. To his right was a silhouette of a farm showing a house, barn, and silo. To my friend, this was the only design that ought to represent her state to the rest of the country.

So, she decided to commit a tiny bit of fraud to tip the balance in favor of Honest Abe.

Illinois officials were holding the voting online, hoping that using this then-new method of citizen engagement would allow them to reach new constituencies. My friend looked at the voting page and realized she could write a simple computer program that would repeatedly vote for Land of Lincoln. It took her all of a few minutes to write the program. She set it to run again and again, stuffing the ballot box in favor of Land of Lincoln. The design won by a landslide. In 2003, the design was launched to the rest of the country.

When my friend first told me this story in 2002, I thought it was funny. I still think about her every time I look at the spare change in my pocket and see an Illinois quarter. At first, I agreed with her that throwing a state quarter election was a harmless prank—but in the ensuing years, I came to see it as sad for the officials. The Illinois officials thought they were getting unprecedented response from the public about a civic issue. What they were really getting was the idle whim of a twenty-something who was bored at work one day. To the Illinois officials, it *looked* exactly like a lot of citizens weighing in on a civic matter. It probably made them happy to imagine that thousands of citizens really, really cared about graphic design on currency. Dozens of other decisions must have been made based on the votes—people's careers, promotions, financial decisions inside the US Treasury.

This is the kind of fraudulent activity that happens every hour of every day on the Internet. The Internet is a magnificent invention, but it has also unleashed an unprecedented amount of fraud and a network of lies that move so fast that the rule of law can hardly keep up. After the 2016 US presidential election, there was a flood of interest in fake news. Nobody in tech was surprised that fake news was out there. They were surprised that people took it seriously. "Since when did people start believing everything on the Internet is true?" one programmer friend asked me. He honestly didn't realize that there are people who don't understand how web pages are made and how they get onto the Internet. Because he didn't realize this, he didn't realize that some people consider reading something on the Internet to be the same as reading something from a legitimate news outlet. It's not the same, but the two *look* so similar nowadays that it's easy to confuse legitimate and illegitimate information if you're not paying careful attention.

Few of us pay careful attention.

This willful blindness on the part of some technology creators is why we need inclusive technology, and we also need investigative journalism to keep the algorithms and their makers accountable. The foxes have been guarding the henhouse since the beginning of the Internet era. In December 2016, the Association for Computing Machinery (ACM), the main professional association for computer scientists, announced it was updating its code of ethics—for the first time since 1992. Many ethical issues have arisen since 1992, but the profession wasn't ready to confront the role that computers played in social justice issues. Fortunately, the new ethics code does recommend that ACM members should address issues of discrimination embedded in computational systems—prompted in part by the efforts of data journalists and academics who have taken on algorithmic accountability.[3]

Consider the case of eighteen-year-old Brisha Borden. She and a friend were goofing around on a suburban street in Florida. They spotted an unlocked Huffy bicycle and a Razor scooter. Both were child-sized. They picked them up and tried to ride. A neighbor called the police. "Borden and her friend were arrested and charged with burglary and petty theft for the items, which were valued at a total of $80," wrote ProPublica's Julia Angwin in her coverage of the event.[4] Angwin then compared Borden's crime to another eighty-dollar infraction: the time that Vernon Prater, 41, shoplifted $86.35 in tools from a Florida Home Depot store. "He had already been convicted of armed robbery and attempted armed robbery, for which he served five years in prison, in addition to another armed robbery charge. Borden had a record, too, but it was for misdemeanors committed when she was a juvenile," Angwin wrote.

Each of these people was given a future risk rating when they were arrested—a move familiar from a movie. Borden, who is black, was rated high risk. Prater, who is white, was rated low risk. The risk algorithm, COMPAS, attempted to measure which detainees are at risk of recidivism, or reoffending. Northpointe, the company that developed COMPAS, is one of many such companies that are trying to use quantitative methods to enhance policing. It's not malicious; most of the companies hire well-intentioned criminologists who believe they are operating within the bounds of data-driven, scientific thinking on criminal behavior. The COMPAS designers and the criminologists who adopted the instrument truly thought they were being fairer by adopting a mathematical formula to evaluate whether someone was likely to commit another crime. "Objective, standardized instruments, rather than subjective judgments alone, are the most effective methods for determining the programming needs that should be targeted for each offender," reads a 2009 COMPAS fact sheet from the California Department of Rehabilitation and Correction.[5]

The problem is, the math doesn't work. "Black defendants were still 77 percent more likely to be pegged as at higher risk of committing a future violent crime and 45 percent more likely to be predicted to commit a future crime of any kind," Angwin writes. ProPublica released the data it used to perform the analysis. This was good because it enhanced transparency; other people could download the data, work with it, and validate ProPublica's results—and they did. This story unleashed a firestorm inside the AI and machine-learning communities. An absolute barrage of debate ensued—of the polite academic form, meaning that people wrote a lot of white papers and posted them online. One very important one was by Jon Kleinberg, a computer-science professor at Cornell University; Cornell graduate student Manish Raghavan; and Harvard economics professor Sendhil Mullainathan. In it, they proved that mathematically, it's impossible for COMPAS to treat white and black defendants fairly. Angwin writes: "A risk score, they found, could either be equally predictive or equally wrong for all races—but not both. The reason was the difference in the frequency with which blacks and whites were charged with new crimes. 'If you have two populations that have unequal base rates,' Kleinberg said, 'then you can't satisfy both definitions of fairness at the same time.'"[6]

In short: algorithms don't work fairly because people embed their unconscious biases into algorithms. Technochauvinism leads people to assume that mathematical formulas embedded in code are somehow better or more just for solving social problems—but that isn't the case.

The COMPAS score is based on a 137-point questionnaire administered to people at the time of arrest. The answers to the questions are fed into a linear equation of the type you solved in high school. Seven *criminogenic needs*, or risk factors, are identified. These include "educational-vocational-financial deficits and achievement skills," "antisocial and procriminal associates," and "familial-marital-dysfunctional relationship." All of these measures are outcomes of poverty. It's positively Kafkaesque.

The fact that nobody at Northpointe thought that the questionnaire or its results might be biased has to do with technochauvinists' unique worldview. The people who believe that math and computation are "more objective" or "fairer" tend to be the kind of people who think that inequality and structural racism can be erased with a keystroke. They imagine that the digital world is different and better than the real world and that by reducing decisions to calculations, we can make the world more rational. When development teams are small, like-minded, and not diverse, this kind of thinking can come to seem normal. However, it doesn't move us toward a more just and equitable world.

I'm not confident that tech's utopians and libertarians are going to make a better world through using more technology. A world in which some things are more convenient, sure—but I don't trust the vision of the future in which everything is digital. It's not just about bias. It's also about breakage. Digital technology works poorly and doesn't last very long.[7] Phone batteries run down and stop holding a charge over time. Laptops stop working when their hard drives fill up after years of use. Automatic faucets don't recognize the motion of my hands. Even the elevator in my apartment building, which should be a simple algorithm based on workhorse technology that was invented decades ago, is flaky. I live in a high-rise building with a single bank of multiple elevators. Every hallway is identical. In one of the elevators, there's something wrong with the wiring or the chips so that every few weeks, you press the button for your floor and it takes you to the floor above or below. It's unpredictable. A number of times, I've gotten off the elevator and walked to what I thought was my apartment door, and discovered that my key didn't work. I was on the wrong floor. The same thing has happened to everyone else who lives in my building. We chat about it in the elevators.

An elevator is a sophisticated machine with some programming embedded in it. There is an algorithm that determines which elevator goes to which floor and which one goes express down to the lobby and which one stops along the way. There are degrees of sophistication in the algorithms, too: newer elevators have programs that optimize the route for the people who push buttons at any given time. In the *New York Times* building, you push your destination floor on a central keypad at the elevator bank, and you're directed to an elevator optimized to get you to your destination as fast as possible, based on other people who also want to get to similar destinations at the same time. However, an elevator has one job. That one job is supported by highly qualified inventors, structural engineers, mechanical engineers, salespeople, marketing people, distributors, repair people, and inspectors. If all these people working together for decades can't make the elevator in my building do its one job, I don't have faith that a similar group of highly skilled people in a different supply chain will be able to make a self-driving car that will do multiple jobs simultaneously without killing me—or killing my kid or killing other people's kids who are riding a school bus or the kids who are innocent bystanders waiting at a bus stop.

The little things like elevators or automatic faucets matter because they are indicators of the functioning of a larger system. Unless the little things work, it's naive to assume the bigger issues will magically work.

Programmers' unconscious biases have been manifest for years. In 2009, Gizmodo reported that HP face-tracking webcams were not recognizing dark-skinned faces. In 2010, Microsoft's Kinect gaming system struggled to recognize dark-skinned users in low-light conditions. When the Apple Watch was released, it did not include a period tracker—the most obvious point of self-quantification for all women. Melinda Gates of the Bill & Melinda Gates Foundation commented on the omission: "I'm not picking on Apple at all, but just to come out with a health app that doesn't track

menstruation? I don't know about you, but I've been having menstruation for half my life, so far. It's just such a blatant error, and it's just an
example of all the things we can leave out for women." Gates also commented on the lack of women in AI research: "When you look in the labs
at who's working on AI, you can find one woman here, and one woman
there. You're not even finding three or four in labs together."[8] Outside of
the lab, the gender balance in leadership positions at tech firms is better—
but not good. According to 2015 diversity figures compiled by the *Wall
Street Journal*, LinkedIn is the big tech firm with the greatest percentage of
women in leadership roles, with a measly 30 percent. Amazon, Facebook,
and Google lag with 24, 23, and 22 percent, respectively. In general, statistics on leadership positions tend to be increased by women who rise to the
top in marketing and human resources. These two departments tend to be
more gender-balanced than engineering roles, as do social media teams.
However, at tech firms the real power is held by the developers and engineers, not by the marketers or HR folks.

It's also worth considering the consequences of sudden, vast wealth on
the community of programmers. Drugs play a large role in Silicon Valley
and thus in the larger tech culture. Drugs were a major part of the 1960s
counterculture, from LSD and marijuana to mushrooms and peyote and
speed. In tech, drugs never became unpopular, but for years nobody really
cared if developers were stoned so long as the code shipped on time. Now,
with the opiate crisis reaching dramatic heights, it raises the question of
how much technologists are facilitating the popularity and distribution of
the ADD drugs and LSD and mushrooms and marijuana and nootropics
and ayahuasca and DIY performance-enhancing drugs that are as popular
in Silicon Valley as elsewhere. "With a booming startup culture cranked up
by fiercely competitive VPs and adrenaline-driven coders, and a tendency
for stressed-out managers to look the other way, illicit drugs and black-
market painkillers have become part of the landscape here in the world's
frothy fountain of tech," wrote Heather Somerville and Patrick May in the
San Jose *Mercury-News* in 2014.[9]

In 2014, California ranked second among all states for highest rate of
illicit-drug dependence and abuse among eighteen- to twenty-five-year-
olds. That same year, the Bay Area had 1.4 million prescriptions for hydro-
codone, a painkiller that is often taken recreationally. If you take speed to
stay up, painkillers are useful to sleep. "There's this workaholism in the

valley, where the ability to work on crash projects at tremendous rates of speed is almost a badge of honor," Steve Albrecht, a San Diego substance-abuse consultant, told the *Mercury News*. "These workers stay up for days and days, and many of them gradually get into meth and coke to keep going. Red Bull and coffee only gets them so far." San Francisco, Marin, and San Mateo Counties had 159 visits per one hundred thousand people to hospital emergency rooms for stimulant abuse. This is triple the national average, which is thirty visits per one hundred thousand people.

Drug use is represented equally across all racial groups, as Michelle Alexander writes in the *New Jim Crow*.[10] However, while poor communities and communities of color are surveilled aggressively to enforce compliance with drug laws, the technology elites who build the surveillance systems seem to be free from scrutiny. Silk Road, an eBay-like marketplace for drugs, flourished openly online from 2011 to 2013. After its founder, Ross Ulbricht, was sentenced to prison, others stepped in to fill the gap. Alex Hern wrote in the *Guardian* in 2014: "DarkMarket, a system aiming to create a decentralised alternative to online drugs marketplace Silk Road, has rebranded as 'OpenBazaar' to improve its image online. OpenBazaar exists as little more than a proof of concept: the plan was sketched out by a group of hackers in Toronto in mid-April, where they won the $20,000 first prize for their idea."[11]

Two years later, an entrepreneur named Brian Hoffman took the Open-Bazaar code, commercialized it, and got a $3 million investment from venture capital firms Union Square Ventures and Andreesen Horowitz to run the marketplace using Bitcoin, an alternative digital currency. In this, we can see the libertarian paradise that Thiel and others imagine: a new space, beyond the reach of government. It seems that their plan is working. Lend Edu, a fintech firm, surveyed millennials about their use of Venmo, a payments app owned by PayPal. Thirty-three percent of respondents said they had used Venmo to buy marijuana, Aderall, cocaine, or other illegal narcotics.[12] A site called Vicemo.com boasts the tagline "See who's buying drugs, booze, and sex on Venmo." It shows a constant livestream of people who publicly post their Venmo transactions. There is very little subtlety. A typical transaction reads like this post: "Kaden paid Cody/for my grub and ganja." Other users post emojis of pills or hypodermic needles. Trees or leaves or the phrase "cutting grass" typically signify marijuana transactions. Some of these are jokes, and some are actual horticultural payments, of course. Either way, it's a little shocking to see the volume of transactions as the country struggles with an opioid crisis.[13]

Illegal drugs are popular. They've always been popular. Most people would argue that illegal drug use is not good, at least not for society as a whole—so when tech is being used to facilitate and distribute them, tech is being used in a way that's counterproductive for cultural good. Yet this is the logical outcome when tech is produced according to libertarian values with a willful disregard for application. If any of those people buying or selling drugs were apprehended and had their stats run through the COM-PAS system, it would perpetuate yet another act of blatant discrimination. Therefore, it isn't enough to ask, of any new technical innovation, if it's good. Instead, we need to ask: Good for whom? We must investigate the wider application and implications of our technical choices and be prepared for the fact that we might not like what we find.

III Working Together

10 On the Startup Bus

Technochauvinists love disruptive innovation. Popularized by Harvard Business School professor Clayton Christensen in his 1997 book *The Innovator's Dilemma, disruptive innovation* is allegedly the technological tidal wave that sweeps away the competition and results in huge profits.

Innovation—and disruption, come to think of it—is usually connected to young people. Ask an executive who he imagines as the ultimate innovator, and odds are he'll paint a picture of a twenty-something computer genius in a hoodie who's writing code to make the next billion-dollar startup. What he means, or perhaps hopes, is that young people can come up with ideas so new, so fresh, so original, that they create an entirely new market: new products to sell, new desire from consumers, a new facet to an existing industry—or a new industry altogether. There's a reason that the *Economist* called disruptive innovation "the most influential business idea of recent years."[1] Just think of all the dollars attached. This hypothetical business executive might also say something about the power of collaboration and how getting creative people together in a room with a whiteboard can create disruptive innovation.

I wanted to see an innovation process from start to finish to find out how much truth there is to these kinds of assumptions. I had a choice: I could join a team in an office for months, helping a team of hackers and business strategists launch a new app or software product into the world—or I could watch the same cycle occur over the course of five days. I chose the latter, which is how I found myself glued to a laptop screen on a ramshackle bus in West Virginia with twenty-seven strangers as part of a wacky computer programming competition called the Startup Bus.

Because many things in Silicon Valley have been gamified, it's no surprise that innovation has also undergone gamification. The best example of this is the innovation contest. Usually, people are motivated to innovate because of conventional financial rewards. If you write a script for a new TV show, the studio pays you to write more scripts and be involved in the show development. There's also *open innovation*, in which people from outside a company develop new tools or products for a variety of altruistic or self-interested reasons.[2] Then there's the *innovation competition*, in which a company announces a challenge and offers a bounty for the best product or solution. The iconic example of this is the DARPA Grand Challenge, the robot car race in which the winner was offered $2 million, and which I wrote about in chapter 8. Incidentally, $1 million is the bounty offered on the game show *Survivor*, which has nothing to do with tech and very little to do with innovation, but which requires surviving elimination challenges on a tropical island with very little food or water, among a group of scheming strangers. For thirty-nine days. On camera.

Would the Startup Bus be like *Survivor*, but with computers and on a moving bus? Or would this particular group of strangers be able to create something new and valuable, something that could shake up the tech industry? I convinced my editor at the *Atlantic* to commission a story so I could find out. At 5:00 a.m., I stood on a street corner in Manhattan's Chinatown with a few dozen techies, ready to innovate my brains out.

On the third day, half the people on the Startup Bus got motion sick. We hadn't slept for two or three nights, the roads through the Smoky Mountains were perilously curved, the tour bus was traveling at top speed, and we had all been staring at our laptop screens for far too long.

Someone on my team bumped the table where we sat and it collapsed on our laps for the third or maybe tenth time that day. Alicia Hurst, my team's designer, grabbed her computer before it fell, but her giant water bottle hit the floor—again. Emma Pinkerton, our business strategist, held up the table while I scrambled to find the bolt that would half-secure it to the wall—again. I dug around under the tangle of backpacks, purses, computer bags, energy bar wrappers, extension cords, and tortilla-chip crumbs until I found the bolt.

Order was restored briefly—then I heard Jennifer Shaw, one of the Startup Bus conductors (as they call themselves) grab the mic. "Hey, hey, hey, New York!" she said, for the hundredth time. Shaw and Edwin Rogers were leading the New York contingent. I was one of twenty-four "buspreneurs" (I know) who had signed up to spend three days on the bus pretending to start a tech company. We were on our way to Nashville, Tennessee, where we would meet up with the four other buses coming from San Francisco, Chicago, Mexico City, and Tampa. All of the teams would compete to determine who had built the best technology company while on the bus.

When Shaw greeted us on the mic, we were supposed to cheer. On the mountain roads, however, she got only weak responses. "That was some weak fucking sauce!" Shaw said. She was thirty-six, relentlessly cheerful, with long red hair and a gap between her front teeth. "Let me hear you! Hello, New York!" The responses were a bit louder this time, and she seemed satisfied. She paused. A brief look of confusion crossed her face, as if she'd forgotten what she stepped up to the mic to do. She had only gotten about two hours of sleep the night before. Rogers, her colleague, took the mic.

"We're going to start pitching again soon," he promised. "We've gone easy on you so far. But the qualifiers are tomorrow, and the judges are not going to be easy on you. They are all entrepreneurs, investors, people who have ridden the bus before. They know how hard it is. You're going to have to show them that your idea has traction. They want to see users, revenue, a product that is going to make a billion dollars." He was getting worked up. His version of coaching involved berating us, as if it weren't already difficult enough being on a crowded bus on which the Wi-Fi only worked sometimes and the broken electrical system meant sharing three plugs among fifty-plus devices. I jammed in my orange foam earplugs and turned back to my laptop, where I was working on a pitch deck for my team's pizza-calculation app—more on what that means in a minute—called Pizzafy. Our website address, purchased on the first day of the trip, was pizzafy.me.

Startup Bus is arguably the looniest of the many hackathons that take place every weekend across the country. A *hackathon* is a marathon computer-programming competition that, among computer programmers, is only slightly less popular than video games, Ultimate Frisbee, or *Game of Thrones*. It lasts twenty-four hours to five days, and usually there's a lot of Red Bull and very little sleep.

Startup Bus is part of a special subset of hackathons, *destination hackathons*, which require attendees to travel to some remote location for the duration. (A spinoff called Starter Island, run by a former buspreneur, requires attendees to spend five days coding on a yacht in the Bahamas.) According to Startup Bus founder Elias Bizannes, some 1,300 people had ridden on one of the buses up to that point and had been "initiated" into the Startup Bus community. The first bus traveled from San Francisco to Austin in 2010, landing its entrepreneurs at the South by Southwest festival. In 2015, the year I rode, buses were meeting up in Nashville at a tech conference called 36|38. June is a better time for a nationwide road trip than March: the year before, the Kansas City bus got iced in on the highway for twelve hours on its way to Austin.

People who haven't participated in hackathons talk about them as hotbeds of innovation, the kinds of environments in which great thinkers come together and dream up exciting new ideas. Hackers don't see it that way. They share an open secret: nothing useful is ever created at a hackathon. There's even a term for the useless software that people make: *vaporware*. The idea is that it's created, and then it evaporates because nobody works on the project after the hackathon (despite everyone's best intentions).

In reality, a hackathon is a sporting and social event. It's like a regatta for nerds. Hackathons also serve as outrageously complex recruiting events. Venture capitalists and head hunters for top tech firms haunt hackathons to spot and poach talent. On the surface, however, nobody talks about hackathon software as ephemeral. People pretend that they're really starting businesses, that they're creating software that will have an impact, that they're doing something that has the potential to change lives. The fantasy of creating the next Google is seductive—so seductive that people sign up to spend days with strangers, forgoing sleep, in order to play at being tech entrepreneurs.

The Startup Bus began as a drunken fantasy. Bizannes, its founder and CEO, had moved to San Francisco in 2010 from Australia, where he had worked as an accountant. He was attracted to startup culture, but he was down to his last couple hundred dollars in his bank account, and if he didn't launch something soon he was going to have to leave California. Drinking with friends one night, he had a brainstorm: What if he launched his own version of Startup Weekend, a popular hackathon—but made everyone ride on a bus? He called Steve Repetti, an investor he knew in Florida, and woke him up. Repetti agreed to invest $5,000 dollars on the condition that he could have a seat on the bus. The project launched a few months later. Bizannes, who went on to work in venture capital at Charles River Ventures, now runs the Startup Bus every year and also runs Startup House, a residential incubator for hackers like the one lampooned on HBO's *Silicon Valley*. He's also infamous for serving as a judge at the 2013 TechCrunch

Disrupt hackathon, in which two participants proposed an app, Titstare, for staring at women's breasts. They later claimed it was a joke, but given what we know about the number and status of women in tech, it didn't read that way. Even as attempted humor, this app should tell you everything you need to know about the true level of disruption and innovation at hackathons. As Betsy Morais wrote in the *New Yorker*: "That misogyny is an institutional reality in a field that prides itself on its forward-thinking worldview makes up only half of the absurdity highlighted by Titstare."[3]

My Startup Bus team was focused on something that I hoped was non-controversial: pizza. My role was somewhat meta; I was on the bus to write about the experience of being on the bus. However, because I'm competitive and good at programming and, I hoped, reasonably innovative, I also wanted to win. I had a plan. It was born of disappointment. At my first hackathon, three years earlier, I pitched an idea for software that I really wanted for myself. It was a community garden finder that would let you enter your location and would list every community garden nearby, along with contact info and the estimated length of the waitlist for a plot.

Nobody wanted to join my garden-finding team.

I learned from that experience that the ideal hackathon project is achievable in the time available, is based on something generically appealing to most people in the room, and has just a hint of whatever is the hot technological topic of the day. Hardware had a turn in the spotlight: for a bit, people were excited about the transformative possibilities of fabrication, sensors, 3-D printing, and wearable technology. Data science was the hot thing for a while, as was artificial intelligence. For this hackathon, I was prepared with a surefire idea. My husband had come up with it as a joke, but the more I thought about it, the more it seemed perfect (unlike Titstare).

When the New York bus left Manhattan at 6:30 a.m. on the first day, an hour and a half after our scheduled departure time, Shaw and Rogers made everyone on the bus stand up and pitch their ideas. Some were more popular than others. Dre Smith, a software developer, got up and said, "My idea is simple. I want to build a virtual-reality dance party." People liked that. Another developer proposed an app to help people schedule conference rooms more efficiently. *That already exists*, I thought to myself. Predictably, there were a couple of ideas for apps that would help millennials meet each other. (Every hackathon includes an idea for making an app that replicates the experience of online social networking in real life.)

A few people on the bus had some variation of, "It's like ____ for ____,"
as in: "My idea is to build an app that's like AirBnB for boats," which was a
pitch by a second red-haired woman named Jen. This seemed profoundly
unrealistic to build in three days on a bus, but I've always wanted to learn
to sail, so I imagined it might be fun to hang out with buspreneurs who
were into boats. I made a note to work on her app if I couldn't form a team
around my own idea.

It was my turn. I stepped up to the mic. "I have an idea for an app that
calculates exactly the amount of pizza you need for a party," I said. People's
heads perked up. "I used to throw a monthly pizza party with a group of
friends, and every time we got hung up on how much pizza to order. We
called it doing pizza math, and we always messed it up. I want to create an
app that calculates the amount of pizza needed for an event based on who
is coming, their age, their gender, and their preferred toppings." People
clapped. I was more relieved than I expected. It was possible that this insane
plan could work.

Our team's pizza technology came together thanks to Eddie Zaneski, the
other hacker in our group. Hackathons tend to be sausage fests, so it was
unusual for our team to be predominantly women. Eddie was twenty-five
years old, six feet seven, and dressed exclusively in free t-shirts from tech
events. "I haven't bought clothes in years," he told me as we sat across
from each other at the rickety table. "I have a bigger wardrobe than my
girlfriend." Eddie was a developer evangelist at a tech company called Send-
Grid, which means his job was to go around the country attending hackathons, throwing pizza parties, and handing out t-shirts to developers to
convince them to use SendGrid. SendGrid is the technology that many
tech companies, including Uber and AirBnB, use to send out autogenerated
emails like receipts and marketing messages. Eddie was worried that he had
brought too many t-shirts. Our bus had only twenty-eight people. Three
giant boxes of shirts, which reached four feet high when stacked, were in
the belly of the bus.

Eddie decided he had more important things to worry about, like getting our pizza-calculator app to work before the hackathon qualifiers in
Nashville, so he put on his blue headphones and turned back to his laptop.
The apple on its lid glowed through a layer of stickers from other tech
events and tech companies: 18F, Penn Apps, GitHub, and HackRU. The latter was Eddie's favorite hackathon: it took place at his alma mater, Rutgers

University. I picked Eddie for my hackathon team because of his stickers. Hackers parse each other's laptop stickers like fashion mavens parse clothing labels. Eddie's sticker from 18F, the government open-data team, suggested that (like me) he was into civic hacking and using technology for social good.

Our app was built using Node.js, a microframework called Express.js, a MongoDB object-relational mapper called Mongoose, and authentication middleware called Passport. We deployed it on Heroku and used Bootstrap for the frontend. These are all free software tools that developers use to make other software. Building an Internet app in 2015 was a lot like building a custom Lego house. The building blocks, or bits of code, were all available on the Internet. The biggest repository is GitHub, a code-sharing site. We decided what we wanted our app to do, grabbed the prebuilt pieces of code that would serve as the structural foundation for the "house," then started building walls and decorating.

Most contemporary software development is a craft, like building houses or furniture. Hackathons are a good way to practice new techniques with other, (slightly) more experienced people in the room. This is another secret in the hacker community. Written instructions and online videos are only useful up to a certain point; to get really good, or to make something really fast, you have to be in the same room with people and you have to talk to them face to face. Legendary information theorist and designer Edward Tufte has a theory about *data density* that explains why face-to-face communication works better than electronic.[4] Tufte writes that human eyes are optimized to take in many, many distinctions within a small area. I can look at the wall behind my desk and see the minute variations in the paint and wall texture, for example. These data points would not show up on a video conference because the video camera, which captures images as pixels in a grid, doesn't have the hardware capacity to capture and show these very small variations. The video would be displayed on a computer screen, which is also constrained by hardware. The screen has a fixed resolution and refresh rate. It only allows your eyes to take in a finite amount of information. By contrast, your optic nerve is taking in yottabytes of information and processing it every moment. You get better information from the high-resolution world. As screens have become higher in resolution, video conferencing has become more popular. However, there is still a hierarchy. An email is effective like a postcard is effective—but a five-minute phone call is better than a two-page email because of the additional texture and complexity and information that you get from intonation and from the sheer fact of connecting and communicating with the person. A high-resolution video conference can be slightly better than a phone call, and a face-to-face meeting is best of all for communicating complex information. However, a low-resolution video conference is worse than a phone call because of the amount of information lost due to pixel distortion and dropped words. Basically, it's about efficiency. In complex knowledge work like computer programming, you can get a lot more personalized, pertinent, data-dense information in a five-minute face-to-face conversation than you would in hours of online tutorials.

The depth of communication is one reason that people go to hackathons. Another reason is the spirit of community. Once we finished all of our communicating and designing, we on the Pizzafy team had to get real people (our friends, our colleagues, and some strangers) to believe in our imaginary company. In order to get "traction" for our app, we had to get real people to sign up for it and also find some kind of marketplace validation. I called up Domino's Pizza, which owns a 9 percent slice of the $40 billion pizza market in the United States. Tim McIntyre, a vice president of communication, was kind enough to take my call. I explained the app: group pizza ordering, algorithm to determine toppings, and so on. "That sounds like a good idea," he said, sounding surprised. "Apps like that—there's a big appetite for them!" Along with fifty-five years of experience, Domino's has online ordering, and even a feature where you can tweet a pizza emoji at them and get your favorite pizza delivered to your door. However, they didn't have a group-event pizza-calculating app. I decided this meant we had discovered an underserved market niche and put McIntyre's quote into the PowerPoint slide deck I was developing for the final presentation.

It's rumored that some programmers make a living by going from hackathon to hackathon and winning. Personally, I have yet to do anything more than break even on prize money. My fellow buspreneurs and I paid $300 each to ride the bus, plus we were paying for our own food and for five nights of quadrupled-up hotel rooms. Chasing a billion-dollar dream is not cheap.

"I basically gave up two months of my life to organize this," Shaw, the conductor, told me during a lunch stop at a Pizza Hut in Punxatawney, Pennsylvania. She was sitting with Mike Caprio, another Startup

Bus alum who was riding the bus to mentor people in code and business strategy.

Shaw offered to pay for lunch. "Thanks for getting this. I am broke as a joke," Caprio said. I was surprised: both Shaw and Caprio had been introduced to me as entrepreneurs who had started two companies. Another secret of the tech community: sometimes *entrepreneur* means "runs a successful company," and sometimes it means "more ideas than money." People inside tech don't talk about money the same way as people in other industries. Hackathoners chat about tech-company valuations like regular people talk about sports statistics. Instacart was the Startup Bus success story: its founders met through the bus and eventually started a company together. Instacart had grown to the point where it was worth $2 billion, as at least a dozen people told me during the trip. Stories like this keep the disruptive innovation myth alive.

By the time the bus arrived at our hotel in Nashville, my team was wrecked: too much junk food, not enough sleep. But our code worked, and we had a presentation written, and we were ready to see what would happen. On the morning of the qualifiers, all the teams piled onto the grimy New York bus to head to the competition location, Studio 615, a warehouse event space in northern Nashville. It was someone's idea of cool: a bright white box with high ceilings, makeshift stage, and club music pumping at top volume. All of the buspreneurs surged in together, piling our laptops onto the long folding tables covered in black plastic. Some people danced. The space was set up like a fashion show, except it was 9:30 a.m., and in the corner was a spread of cinnamon pecan rolls and sweet tea. Eddie's t-shirts were piled chin-high on a table, along with free t-shirts from two other tech companies and several boxes of stickers.

The first round of competition took place in the green room, a small retreat off the main warehouse space that was covered in charcoal-gray geometric wallpaper. On the wall hung a floor-to-ceiling painting of a naked woman lying in a desert at sunset, holding a can of Reddi Wip.

The judges crammed on a couch to watch each team pitch: Bizannes, Repetti, and the two Startup Bus national directors, Ricky Robinett and Cole Worley. The national directors were the only ones getting paid. Everyone else was a volunteer, including the conductors. I had been warned that the judges would ask about monetization, or how the proposed company would make money. The first team, Shar.ed, shuffled into the green room, plugged in a laptop, and prepared to pitch. I had been next to them on a bus for three days, but wasn't sure what their project was. Shar.ed pitched an idea for crowdsourced, on-demand, for-profit education, in which people could vote for classes they wanted to take and instructors would prepare exactly those classes. They had started an Indiegogo crowdfunding campaign to partially fund the project and had already collected a few hundred dollars.

Next up was Screet, a service that proposed to deliver on-demand products to couples in the throes of passion. Aimed at those who want to be safe but don't want to make a trip to the drugstore, Screet was a smartphone app that would summon a Lyft or Uber driver to unobtrusively drop off condoms, dental dams, or latex gloves that he or she kept stored in the car trunk in plain, SKU-labeled boxes. This service was going to be especially useful for LGBTQIA people, Screet claimed, because dental dams are hard to find in stores. After two pitches, I wandered out of the green room and rejoined my team to watch the rest of the pitches on the simulcast. Pizzafy was next to last.

I was nervous. I pitched. We made it to the semifinals! So did Screet, along with a Chicago team that had made a toy that was controlled by an iPad app and that helped kids and parents play imaginary dinosaur games together, along with some other teams.

We ate boxed lunches. The music played. We pitched again, on the small stage in the main space this time. The pitches were livestreamed. At least a dozen people from other Startup Buses tuned in online. SPACES, a New York team that was working on a virtual-reality app, got up on stage and thanked the judges. "We are grateful for the opportunity, but our presentation contains proprietary material, and we are going to decline to pitch," said John Clinkenbeard, the team CEO. The room erupted. Edwin Rogers started whooping, "New York bus! New York bus! New York bus!" The team, which included Dre Smith of the virtual dance party, had secured $25,000 in funding from an outside investor. The team came down from the stage, then went through the crowd shaking hands and accepting congratulatory hugs and enjoying the ruckus. The national directors, standing to the side in their headsets, looked angry—as if SPACES had transgressed by bringing the stage show to a premature climax. It was a hard act to follow.

Pizzafy advanced again, along with Screet, plus an education company from the Mexico City bus and a Chicago bus project that sends a text message when you take a pill. Everybody else went out to party in Nashville that night. Emma, Eddie, Alicia, and I went back to the hotel. People from other buses came in from drinking and sat down to help or chat. People talked about their lives outside the bus. It started to feel like I imagine a barn-raising feels: lots of people from the community showing up and helping to make something that will benefit only a few people, because everybody eventually needs a barn. These hustlers, hackers, and hipsters on the bus would eventually need to hire people or hire companies or get a breathtakingly specific technical question answered in the real world, and they were laying the foundation for this by helping to build our pizza-party app. This is the other secret to hacker culture: sometimes you do a lot of insane technical work that has no apparent purpose. You do it because it's a rush, like marathons.

We worked all night and all of the next day. We designed an audience-participation stunt, redid our slide deck, honed the pitch until I had every pause and every pizza pun memorized. Finally, in the late afternoon it was time for the final pitch. One of the Instacart guys was a judge for the final round. I got onstage and pitched my best.

The runner-up was PillyPod, "a device that alerts you when loved ones don't take their meds." They had started the week with the URL pillypad. co, but discovered that pillypad.com was an adult-content site, so they changed the vowel in their name.

Then it was time to announce the winner, and I heard the judge say the name of our team. The club lights started going mad, and the DJ blared Katy Perry's "Dark Horse." The four of us made our way up to the stage. Shaw hugged me, Rogers hugged me, people I didn't know hugged me. Rogers cried. I stood on stage with my team and for a few minutes I felt amazing.

After my week on the Startup Bus, I can tell you what it feels like to win a hackathon. It feels like gorging yourself at a pie-eating contest and discovering that the prize is … more pie. Yes, I'd be happy to sell my new pizza-technology company for a bunch of money. I'm not holding my breath. One big secret of hacker culture is that overnight success is a black swan, a lightning strike, an outlier. Useful, lasting technology can't be built in a weekend, or even a week. It's a marathon, not a sprint.

We have cultural fantasies of what can be achieved with computation at hackathons. The reality, however, is often quite different than what we imagine. The Startup Bus is a reminder that there is a lot of hyperbole about the transformative possibilities of technology.[5] None of the apps created while I was on the New York bus made it big. The SPACES team, which got funding during the event, fell apart soon afterward. In reality, software is rarely disruptive or innovative, and even more rarely is it both together (with notable examples, such as Google Search, excepted). You'll remember that our pizza app was almost completely made from pieces of other people's code and that calculating how many pies to order is not disrupting or innovating anything. It's just automating a calculation that was previously done by hand. Yet I learned some things from being on that bus: I became better at coding, I became better at pitching, and I made some contacts that may come in handy for a future project.

Software development is primarily a craft, and like any other craft— woodworking, glassblowing—it takes a very long time (and a period of apprenticeship) to achieve competence at it. Engaging in and democratizing that development work might not look or sound as sexy as coming up with some earth-shattering tech idea, but it's where the future lies.

11 Third-Wave AI

In the book so far, we've looked at why AI doesn't work as well as we might expect. We've looked at miscommunication, at racism disguised as predictive analytics, and at shattered dreams. It's time to talk about something more cheerful: a collaborative path forward that pairs the best of human effort with the best of machine effort. Humans plus machines outperform humans alone or machines alone.

We'll start with a story about teenaged me, and my lawn. My parents' house was a former farmhouse, set on about an acre of land. Starting at age eleven, it was my job to cut the grass. We had a small riding mower. I thought this was terrific; it felt like one step away from driving a car. Like many suburban kids, I couldn't wait to get my driver's license. In the nice weather, every Saturday found me zipping around the yard cutting the grass on the riding mower. I didn't like cutting the grass, but I *really* liked driving the riding mower.

It was an old house and a big yard and built on a hill, so the landscaping was relatively complex. I had to mow an irregularly shaped wide-open space in the back of the house, two formal circular gardens on each side of the house, and a J-shaped patch in the front.

I had a circuit that I drove around the landscaping. I started in the back yard and did a lap around the perimeter of the wide-open space to get the edges. The wheels of the tractor left parallel marks around the track. Then, in the next lap, I drove the tractor so that the right front wheel went exactly in the track made by the left wheel on the previous lap. This ensured that the blades cut even rows, and when I looked out the window at the yard after I was done, it made a sort of deconstructed spiral pattern that I liked.

My mother, an avid gardener, had designed intricate gardens in the yard's different microclimates. A few of these gardens featured sharp ninety-degree corners. It looked elegant. However, the turning radius and the blade placement of my riding mower meant that I couldn't cut a ninety-degree corner without driving four feet into the flowerbed. I could cut an arc close to the corner of the flowerbed, but it was a curve, not an angle.

I could have done most of the job with the riding mower, and then I could have gone back and finished the corners with a hand mower so that they were right-angled, not curved. Ralph, the guy who cut the grass before I was eleven, did this. In fact, he did the whole lawn with a push mower. Had I been a better person or a better daughter, I would have done it. My mother asked me to do it about a million times. I almost never did it. I could give some excuses (allergies, exhaustion, heatstroke), but I suspect the real reason was that I was a stubborn kid who simply didn't want to because I found it unpleasant. I hated the way the grass and sticks blew out of the hand mower and hit me in the legs so that I got welts and hives. I hated the gasoline fumes and the waves of heat that came off the hand mower. I hated that I felt like I was choking the whole time I pushed the hand mower because I'm allergic to grass. On the riding mower, I was above and in front of the grass-discharge spot. With the hand mower, I was directly behind the grass-discharge spot. The hand mower made me miserable.

My mother eventually gave up and redid the flowerbeds so they were curved instead of angled.

That riding mower is like a computer. My parents bought the riding mower because it was supposed to be a labor-saving device. Instead of hiring Ralph to mow the lawn, they could "hire" me to do the same job at a reduced price. However, the riding mower (which I piloted on the same track every single time, much like a Roomba automated vacuum travels around a room) was built differently than Ralph's hand mower. It didn't do the job the same way. Also, the staff was different. Ralph, a professional landscaper, did a professional job. I, the sullen daughter with grass allergies, did an unprofessional job. My mother was forced to decide: Did she want the inexpensive option that used the fancier technology that didn't do the job she wanted? Or did she want the more expensive option that used the less-fancy technology that did exactly the job she wanted?

My mother was a practical woman who had a lot of children and a lot of gardens, so she opted to make curvilinear garden beds. This is what we end up doing a lot of the time when it comes to automation technology. Automation will handle a lot of the mundane work; it won't handle the edge cases. The edge cases require hand curation. You need to build in human effort for the edge cases, or they won't get done.

It's also important not to *expect* that the technology will take care of the edge cases. Effective, human-centered design requires the engineer to acknowledge that sometimes, you'll have to finish the job by hand if you want it done. An automated phone system will take care of most ordinary problems faced by people calling an airline, for example—but there will always need to be a person answering the phone because there will always be exceptional cases. Likewise, in a newsroom, automation can be helpful for a vast number of things—but there always needs to be someone answering the phones or looking at the automatically generated stories before they're published because technology has its limitations. There are things that a human can see that a machine can't.

There's a name for systems like this that include humans: *human-in-the-loop systems*. For the past few years, I've been interested in building technology out of this framework.[1] In 2014, I was looking around for a new AI project, and I asked a handful of journalists and programmers where they thought the next hot area was going to be. Campaign finance was the overwhelming vote. The US presidential election was coming up; *Citizens United*, the 2010 decision that opened the floodgates to super PAC spending, had made a huge impact on political fundraising. Data reporters were on top of campaign finance.

I decided to join the fray. I put together a plan for a new artificial intelligence engine to detect campaign finance fraud and investigate privacy. It's a human-in-the-loop system that automates the process of discovering new investigative story ideas. Like many AI projects, it works beautifully, but it also doesn't work well. Looking at how the project was built is a way of gaining insight into why AI is wonderful and useless at the same time.

Some investigative stories are like shooting fish in a barrel, and these are the perfect stories on which to deploy AI. To use a computer to find a story, you first need to be fairly sure there's a story to be found. Stories abound in pools of money. Whenever you have a big pool of money, there's inevitably someone trying to steal it. Hurricane recovery, economic stimulus packages, no-bid contracts: if you want to find someone up to no good, it's usually easy to find them lurking wherever there's a lot of cash.

The big pile of money that is federal political campaigns always attracts a few bad actors. In general, politicians are excellent stewards of public funds and are devoted public servants. Sometimes, they are not. In general, the sentiment among data journalists was that we should keep a close eye on campaign funding in advance of the 2016 presidential election.

I had built AI software for the textbook project I mentioned in chapter 5. I was curious to see if I could apply the software, which I called a Story Discovery Engine, to a different context. In tech, we talk a lot about the value of iteration—building something, then rebuilding it even better. I wanted to iterate. I had built a tool that visualized problem sites in one school district. Could I build a tool that would visualize problem sites in the ultimate district, Washington, DC?

With the generous support of a grant from the Tow Center for Digital Journalism at Columbia Journalism School, I decided to develop a new Story Discovery Engine focused on campaign finance. The tool would let reporters quickly and efficiently uncover new investigative story ideas in campaign finance data. The previous engine was built around the idea of helping reporters to write one story—about books in schools. This time, I wanted to build something that would help reporters find a wide variety of stories in a general topic area. I wanted to build a bigger system, and I wanted it to automate more of the grunt work of investigative journalism. It was well in advance of the election, so there would be plenty of time to build the tech and roll it out and use it for reporting on the election.

I had heard a lot about dark money and super PACs in the wake of the 2010 *Citizens United* decision, but I knew there was a vast amount I didn't understand about this complex system. Like public education, campaign finance was a complex bureaucratic system with abundant data. It was a good test case to develop a new engine against. I wondered: Could I identify lawmakers who weren't following their own rules?

I started with a design-thinking approach. In other words, I talked to people who knew a lot about the thing I wanted to do and was guided by what they said their world was like. I talked to experienced reporters and campaign finance experts. I interviewed a diverse range of campaign finance data experts: journalists, Federal Election Commission (FEC) officials, lawyers, people who run campaign finance watchdog groups. Particularly helpful were the designers and developers who worked for 18F, the government's rapid-response technology team.

While I was building my tool, 18F was building a new user interface for the outdated FEC website. FEC.gov is the primary distribution channel for all US campaign finance data. It's traditionally been difficult to navigate and thus difficult to understand. The new interface, rolled out gradually, made information more visible. However, it didn't surface exactly the information that journalists would need to find stories. Instead, it was focused on simple and effective distribution of FEC data (a noble goal). Therefore, I focused my efforts on designing an interface that would do for journalists what 18F's new website didn't. My most essential informant was Derek Willis at ProPublica, a journalist who (probably) knows more about campaign finance data than people who work at the FEC. Willis, who has been reporting on campaign finance for decades, has created a full slate of helpful automated tools for campaign reporting: OpenElections, Politwoops, and others. His work is so good that there's no point in redoing it. I wanted to make something in the margins, something that would add to the discovery tools (like Willis's) out there but would make the reporting process faster. In addition, I read. The most challenging part was reading hundreds of pages of US Code and FEC legislation and policies. I took notes on the common themes that emerged. I paid close attention to the vocabulary that people used.

The first step was designing the architecture of the system. Software has underlying architecture, just like buildings do. The Story Discovery Engine is an AI system, but it doesn't rely on machine learning. It comes from a different branch of AI programs called expert systems. The original idea, back in the 1980s, was that an expert system would be like an expert in a box. You would ask the box a question, like you ask a question to a doctor or a lawyer, and the box would give you an informed answer. Expert systems never worked, unfortunately. Human expertise is too complex to be represented in a simple binary system (which is what computers are). However, I decided to hack the expert system idea, and turn it into a human-in-the-loop system that ran based on rules taken from reporters' subject matter expertise. It worked well. I didn't make a box that told me the answers, but I did make an engine that helped me as a reporter to find stories faster.

I decided that the rules of the new Story Discovery Engine would be determined by the rules of the real-world political system. This was both a smart decision—because I wouldn't have to create new computational rules myself—and a flawed one because the rules for campaign finance in the United States are of Talmudic complexity. I'll attempt to cover them briefly: Each candidate for federal office has an authorized committee.

Individual citizens are limited in the amount they can give to individual candidates through an authorized committee; this limit is currently $2,700 per election. Other political action committees (PACs) may fundraise and may donate to candidates' committees. There are also limits to what PACs may say and may give. Super PACs, or independent expenditure-only committees, may raise and spend unlimited funds on behalf of a candidate. However, they may not coordinate such spending with the candidate or the candidate's official committee. Other groups of interest are leadership PACs, Carey committees, joint fundraising committees, 527s, and 501(c)s, all of which collect, spend, or engage in electioneering on behalf of or in opposition to one or more candidates. Committees and PACs are required to report their expenditures and receipts to the FEC. 527s and 501(c)s are required to report to the Internal Revenue Service (IRS).

Say what you will about American government bureaucracy, but it truly is well-suited to database modeling. Bureaucracy is a byzantine maze of rules and regulations, carefully spelled out. Fraud, or at least shenanigans, happen in the cracks between the rules; computer code is a giant set of rules. Therefore, if we get a little creative in how we express the rules computationally, we can efficiently model how things are supposed to work in campaign finance. Then, we can figure out where to find things that went wrong. I put together a diagram that modeled the entities and the relationships between them. The entities became objects.

Campaign finance fraud is a useful phrase; it's only the top level, however. There's actually very little fraud in campaign finance because very little is illegal anymore. In the 1970s, the United States put in place very strict limits on how much candidates could raise and spend and from whom they could raise money. Landmark decisions since then have rolled back those limits. In 2002, the Bipartisan Campaign Reform Act made it so that the limits on contributions to federal candidates and political parties increase every few years. The 2010 *Citizens United* decision made it so that outside groups like super PACs can raise and spend unlimited funds on behalf of candidates, so long as the super PACs don't coordinate these expenditures with candidates. Another 2010 decision, *Speechnow.org v. FEC*, removed restrictions on what outside groups like 527s can raise. These groups now merely need to disclose their donors. In 2014, *McCutcheon v. FEC* removed the upper limit on what an individual can give to candidates, political parties, and PACs combined.[2] A full explanation of campaign finance is beyond the scope of this book, but I highly recommend reading the website of the Center for

Responsive Politics, which offers an excellent primer on what laypeople need to know about campaign finance.

After I talked to all the experts, I extracted the common elements from the conversations. All the experts had certain types of anomalies that they looked for when they were looking for (or at) campaign finance shenanigans, certain red flags that kept coming up over and over again. Administrative overspending was one such red flag. To understand administrative overspending, we need to start with a definition: All political committees are technically nonprofit corporations. Unlike regular nonprofits, however, political committees file financial reports with the FEC rather than the IRS. At every nonprofit corporation, some of the organization's money is spent on its purpose, and some is spent on keeping the organization running. The purpose-driven expenses are called *program expenses*. The internal expenses are called *administrative expenses*. At a political committee, program expenses might be electioneering expenses: the costs of buying television, print, and digital ads; the costs of purchasing yard signs; or donations to political candidates. Administrative expenses are expenses like salaries, office supplies, or the costs of organizing a fundraiser. The ratio of administrative expenses to overall expenses is a measure of health for any nonprofit. Many people use this ratio to evaluate whether the nonprofit is well-run when they're deciding who to donate to.

Another example of something to look for is a vendor network. Let's say that candidate Jane Doe is running for president. Supporter Joe Biggs wants to give a million dollars to Doe's cause. Donated money doesn't go directly to the candidate, remember; it goes to the candidate's primary campaign committee, Jane Doe for President (JDP). However, Biggs can't give a million dollars to Doe's campaign committee; the personal donation limit is $2,700. However, Biggs is welcome to give that million dollars to a super PAC, Justice and Democracy Political Action Committee (JDPAC), which can spend the money however it likes in order to get Doe elected. JDPAC spends Biggs's money on what are called independent expenditures. The catch with an independent expenditure–only group (like a super PAC) is that it can't coordinate with the official campaign committee at all—so JDPAC is prohibited from coordinating its efforts with JDP.

Now, let's say that JDP hires a graphic design firm in Wichita to create its campaign ads. That firm's name, Wichita Design, will show up in JDP's spending reports filed with the FEC. Let's also say that JDPAC happens to hire the same graphic design firm. This will also show up in JDPAC's spending reports filed with the FEC. It's possible that there's no coordination. The graphic design firm could have really good industrial hygiene: it could set up a firewall internally and educate its staff that there isn't supposed to be any coordination, and it could do a good job of keeping the two accounts separate. This is entirely possible, and it's legal and appropriate. It's also the case that many, many committees use the same vendors for ordinary tasks. There are only a limited number of payroll-processing companies in the United States, for example. Most campaigns and outside groups use ADP for their payrolls, and it's not a story. However, it's equally possible that there's coordination happening at the vendor level. Therefore, if a journalist can easily see that JDPAC and JDP are using the same graphic design firm in Wichita, which happens to be run by Jane Doe's college roommate, the journalist definitely is going to follow up and see if there is illicit coordination happening. It's quite likely to be a story.

It's traditional to give a software project a name, like naming a pet. It provides a shared reference point that the people on the project can use. I decided to name my project Bailiwick. *Bailiwick* has two definitions, according to Merriam-Webster: "the office or jurisdiction of a bailiff" or "a special domain." Both definitions seemed to fit, especially since a *bailiff* is "an officer in a court of law who helps the judge control the people in the courtroom." I imagined that my program would fulfill the role of the tall, bald bailiff named Bull or the wisecracking bailiff Roz on the 1980s TV show *Night Court*. It would carry documents back and forth between the people and the data, and it would serve a quasi-official function as an intermediary. I also liked that the word *bailiwick* sounded kind of cute and playful. In my world, anything that can make campaign finance data more playful is entirely welcome.

On a more practical note, a software application must have a name because you have to put it into a directory on your computer, and that directory has to have a name. It's essential to pick the name at the beginning, just as with a baby. On the other hand, if you name your baby Joseph and decide two days later that you want to call your baby Yossi, you just start calling the baby Yossi and write "Yossi" inside his t-shirts. With a computer program, if you change the name of the base directory, you can create major headaches inside your code.

So, Bailiwick it was. Bailiwick can be found online at campaign-finance .org.

Which brings us to the development process. Some of the difficulties I faced during the project are emblematic of the kinds of ordinary challenges that arise during any coding project. For example, I decided to hire someone to help me code because my deadline was tight. Hiring a developer is not unlike hiring a lawyer: The good ones are insanely expensive. They are also hard to find, because they don't advertise. They don't need to. There are a few directories, sure, but for the average person it's quite difficult. I searched online for "hire Django developer." I got a big pile of garbage. Here's a sample of one of the search results:

Django Jobs | **Django developers** | Freelance Jobs
Django team is one of the most popular **django** freelance jobsite in online. **Django** team is a souk for top **django developers**, engineers, programmers, coders, architects ...

Searching online for a developer was just too hard. Instead, I found myself working my personal networks for recommendations. Online hiring for professional services is an example of an area in which technology was supposed to make things easier, but it actually made things harder. The algorithmic layer on top, which can be manipulated for profit, interferes with the average individual's ability to do something simple like find a software developer. The same problem arose when I tried to find a handyman to fix something in my house. It reminded me why curation is so useful. In an online world in which everyone is supposed to find their own truth, it can sometimes take forever to do simple things. The paradox of choice can be a burden.

Regrettably, I found myself in the same position as the nineteenth-century mathematicians who needed more human computers and couldn't find them. I wanted to hire an entire team of women and people of color. I worked all my networks; it was far more difficult than I anticipated. I talked with a developer who runs her own shop who is a woman of color; I couldn't afford her. Nor could I afford the heavily discounted services offered by a friend's software firm. Eventually, I hired a woman and three men, all independent contractors, bringing the project total to a 2:3 women:men ratio. On a small team with a very close deadline, that would have to do.

It's an open secret in project management that nobody knows how to estimate time for a software project. Part of the problem is that writing computer code is more like writing an essay than like manufacturing. Original code hasn't been written before, so there's not really a good way to estimate how long it will take to make it—especially if the code is intended to do something that hasn't been done before. Another problem is that people, not machines, write code. People are bad at estimating time and effort: they go on vacation; they spend the afternoon messing around on Facebook instead of programming. In short, they are people. They are variables, not constants.

Representing the complex relationships of the campaign finance world in a simple, easy-to-discover manner was a challenge. I worked with a user-interface expert, Andrew Harvard, who designed a set of pages that allowed reporters to efficiently organize and sort through the information that mattered to them. State reporters generally care about finding stories relevant to races in their states. National reporters generally focus on the presidential race plus key state races. Regardless, the system lets you select which races and candidates you care about. These are shown in the favorites list when you log in. Figure 11.1 shows what a reporter sees if she has favorited 2016 presidential candidates Hillary Clinton, Donald Trump, and Bernie Sanders. Clicking one of the names takes you to a page that shows a candidate. Each candidate files a series of financial reports with the FEC. A reporter can use

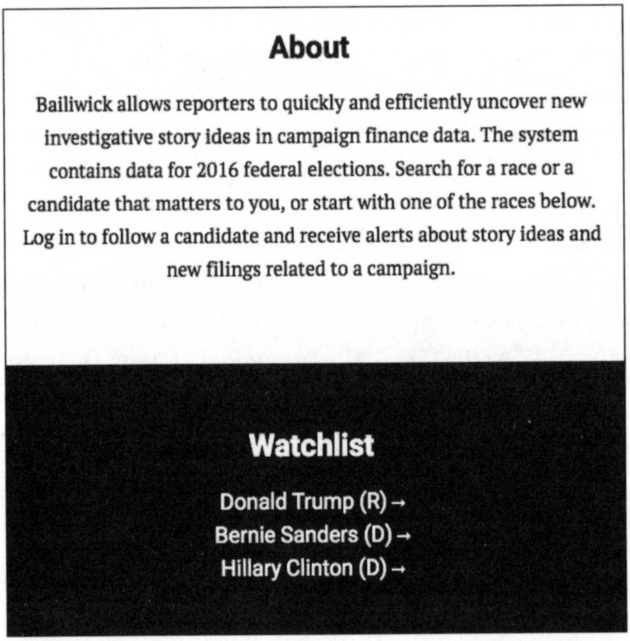

About

Bailiwick allows reporters to quickly and efficiently uncover new investigative story ideas in campaign finance data. The system contains data for 2016 federal elections. Search for a race or a candidate that matters to you, or start with one of the races below. Log in to follow a candidate and receive alerts about story ideas and new filings related to a campaign.

Watchlist

Donald Trump (R) →
Bernie Sanders (D) →
Hillary Clinton (D) →

Figure 11.1
Bailiwick splash screen customized to show three 2016 US presidential candidates.

Bailiwick to scroll through and read the individual financial reports or see the financial report totals organized in a convenient way.

We tend to think of donations as being for and against. However, the campaign finance laws divide the donations into groups. Remember that there are authorized donations and independent expenditures? Bailiwick parses these reports and organizes them into supporting and opposing groups. This saves time and effort. It's also easy to skim through and see relevant names.

The inside and outside groups form the substance of a tree map, a common type of data visualization, that appears at the bottom of each candidate's page. It's hard to parse numbers; it's far easier to see patterns in the data when each expenditure is grouped into categories. In a treemap, each category is a rectangle. The relative size of each rectangle matters, as does the number of donors and the total donation amount. I can click any rectangle to see more detail. As of the inauguration, Great America PAC was the group that spent the most in independent expenditures (see figure 11.2).

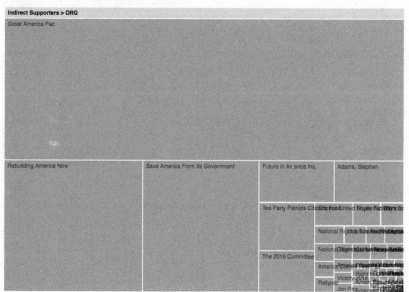

Figure 11.2
Independent expenditures in support of Donald Trump.

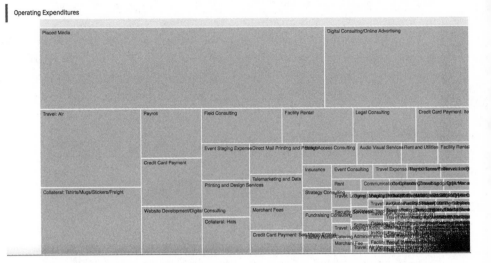

Figure 11.3
Operating expenditures for Donald Trump campaign committee as of December 2017, organized by category. Note the rectangle at the bottom labeled "Collateral: Hats."

Clicking in, we can see that this donor spent $12.7 million dollars in support of Trump's campaign, in dozens of separate transactions over the course of the race.

Data visualizations often trigger story ideas. For example, the first time I saw the tree map of Trump's campaign committee spending pattern, I saw there was a fairly large rectangle devoted to hats (see figure 11.3). As of December 2016, the campaign had spent $2.2 million on hats from a company called Cali-Fame (see figure 11.4).

I didn't know anything about Cali-Fame in the fall of 2016, but it seemed to me like there might be a story in Trump's spending on hats. Reporter Philip Bump had the same idea. On October 25, 2016, he published a story in the *Washington Post* titled "Donald Trump's Campaign Has Spent More on Hats than on Polling."[3] Not only that, but the Trump campaign also spent $14.3 million on t-shirts, mugs, stickers, and freight, all with a single company, Ace Specialties LLC, which specializes in workwear for the oil and gas industry. The company owner, Christl Mahfouz, is on the board of the Eric Trump Foundation.[4] Does this mean anything? I don't know. If I were a reporter on the political beat, that would be another story idea to run down.

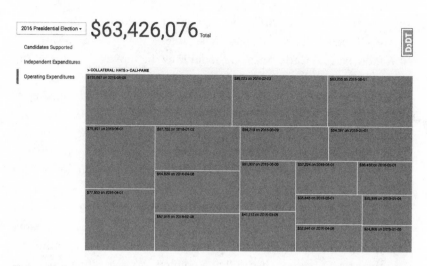

Figure 11.4
Payments from Donald Trump's campaign committee to Cali-Fame marked "Hats," organized by date and amount.

Andrew Sheivachman, a reporter for a travel-industry site called Skift, had a different perspective. He used the tool to develop a story called "Clinton vs. Trump: Where Presidential Candidates Spend Their Travel Dollars." In it, he analyzed how Trump was using campaign funds to pay his own company, TAG Air, for campaign travel.[5] This is not illegal, but it is notable. It's also an opportunity to talk about the many things in campaign finance that are legal, but perhaps not appropriate. The only way we're going to launch a public conversation about these issues is by telling stories. Telling stories is how we understand the world. There aren't easy answers. We need a public conversation, a conversation that includes diverse voices, to resolve these questions in a democratic manner.

The Story Discovery Engine is a human-in-the-loop system rather than an autonomous system. The difference between a human-in-the-loop system and an autonomous system is like the difference between a drone and a jet pack. The difference matters for effective software design. If you expect the computer to do magical things, you'll be disappointed. If you expect it to speed things up, you'll be fine. This attitude of preferring humans assisted by machines is catching on in the $2.9 trillion US hedge fund business, which has always been on the cutting edge of using quantitative methods. Billionaire Paul Tudor Jones, head of Tudor Investment Corp, famously told his hedge fund team in 2016: "No man is better than a machine, and no machine is better than a man with a machine."[6]

A more general way of thinking about how the tool works is that it surfaces the difference between *what is* and *what should be*. *What should be* is that a group's administrative expenses should be less than or equal to 20 percent of its total expenses. *What is*, is whatever percentage of the annual expenses are categorized as administrative according to financial-reporting documents filed with the FEC. If there's an anomaly—if the administrative expenses are greater than 20 percent—then there's an opportunity for a story.

Note that I say *opportunity*. There isn't definitely a story, because there can be a perfectly good reason for having a large amount of administrative expenses in any given quarter. We don't want to create a machine that says there's a 47 percent chance that a political group is acting unlawfully because its administrative expenses are 2 percent higher this month than last month. That would be absurd—and possibly libelous.

Often, when I talk to computer scientists, they suggest looking at the five highest results, the five lowest results, and the average values in a dataset. This is a good instinct, but it isn't always interesting from a journalistic perspective. Let's say that we pull a list of salaries for employees of a school district. The five highest-paid employees are likely to be the superintendent and the highest-ranking executives. The five lowest-paid employees are likely to be nonunionized, part-time employees. This isn't news. It might be surprising or mildly interesting to someone who hasn't seen a lot of salary scales, but that's different than being newsworthy. In journalism, we have an obligation to be both accurate and interesting to a mass audience. Computer scientists have the liberty to be interesting on a smaller scale to a highly trained audience (which is something that makes me eternally jealous). The threshold for interestingness is radically different in each field.

If I was going to look at groups that had large administrative expenses, I would probably look at the ones that had really large administrative expense percentages first. The outliers are the low-hanging fruit. I would look at the groups with the highest percentages and the lowest percentages, and I would see if there was something interesting there.

I made one major modification to the Story Discovery Engine. When I tried to explain the textbook engine, people often asked me, "You mean

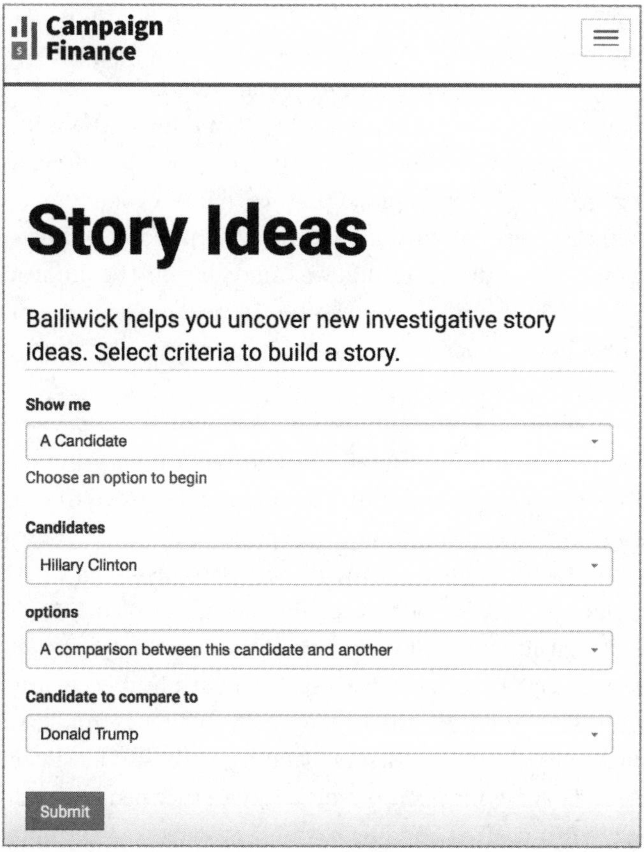

Figure 11.5
Story idea page.

that you made a machine that spits out story ideas?" I explained that it wasn't a machine that spit out story ideas, that it was subtler, and I talked about automation. Most people's eyes glazed over at that point. So, for the second Story Discovery Engine, I decided I would try to make an actual machine that spit out story ideas. Figure 11.5 shows what the feature looks like.

I should specify that unlike other features, the story ideas feature is a minimum viable product (MVP). It works, and you can see an actual result—but only for one case, not for all the cases that we planned. We say this very specifically in the documentation. It works well enough for me to feel confident claiming that it works; from my perspective as a developer, it's a solved problem. But in software, things can work without really working *well*. It's not a binary situation. A person can't be a little bit pregnant, but a software program can be a little bit functional. The point of an MVP is to get the product working well enough that you can demonstrate it to people and get customers or get funding for the next round of development. This isn't good design, nor is it good practice, nor is it good for users to have half-functional pieces of software out there in the world. However, it's evolved to be standard practice. I think we can do better. The problem, most of the time, is the problem I ran into with Bailiwick: the team ran out of money, and thus development time, before we could finish the story ideas feature.

Here's another example of a problem that's quite typical for the development process, but can have wide-ranging effects unless caught. One day, my code was throwing an error that I didn't understand. I decided to make a new database and test the code by loading in all my 3.5 million records again from scratch. It worked great for the first ten seconds—then I got a different error. I fixed something that I thought was the problem. Then, I tried to load the data again. It didn't work. I changed something else that I thought was the problem. That made it worse. I switched back to the first database and tried to recreate the error. I got a totally different error. I realized I wasn't going to be able to fix the first database—ever—so I switched over to the second database for good. I felt bad; the other people on my team were using the first database, and by having it available but corrupted I was getting in the way of other people's ability to code. It was a mundane version-control problem, but because precision matters in computation, the errors I caused were probably creating a cascade of other inexplicable, frustrating errors for other people.

These are the kinds of obstacles that get in the way of adopting technology in the newsroom. By working through the obstacles on a small scale, it's possible to see how a large-scale effort could work. It's also possible to see why the large-scale effort might be failing. We can also see why writing code is not the kind of thing that can be accomplished in an assembly line. There's the factory model—with the assembly line—and the small-batch model. In a factory, you look at all the tasks and decide which ones can be automated and are repeatable. In small-batch production, you do the same thing—but you still do some stuff by hand. Think of computational journalism as being like the slow-food movement.

Thus far, the impact of the tool has been small but mighty. I don't track how many reporters use the tool for stories, but I do use the tool regularly in my classes. I teach about thirty students every semester. This means that at least sixty stories a year can be produced out of the tool. This isn't bad for what it cost to develop the tool. If the tool were used regularly in a newsroom, each story written out of it could generate revenue via the advertising placed next to it on the page. It's never going to be a blockbuster source of revenue, but it would be a drop in the bucket. The tool wouldn't generate as much money as mass-produced assembly-line products, but it would be a revenue-generating, artisanally produced product.

For now, my campaign finance tool doesn't generate any money. In financial terms, it doesn't have a path to sustainability. Bailiwick has value as a teaching tool, as a model for investigative projects, and as an example of applied research (meaning "the opposite of theoretical research") in computational journalism. Much to my chagrin, that intangible value doesn't help with the $1,000 a month it costs to keep Bailiwick's servers running. This is another secret of the tech world: innovation is expensive. If I'd known that the project would cost this much, I might have made different choices along the way—but because nobody had created this type of software before, there was literally no way to forecast expenses. I had a kind of blind spot when it came to the project's operating expenses. This is the kind of blind spot that almost always happens when you build new technology: you need to have faith that you can invent the thing you're trying to make and have faith that the financials will work out. Engineering is sometimes a thrilling leap into the unknown.

12 Aging Computers

I do not plan to run Bailiwick forever. At some point, I will take it offline, archive it, and move on to another software project. Like a car or a plant or a relationship, software needs care and constant attention. It also has a lifespan.

Websites and apps and programs always break because the computers that they are on wear out and need to be updated. The world changes. Software needs to be updated. When you host even a simple website with a company, that company will always go through management changes or be sold or upgrade its servers, and something will inevitably be screwed up. Every year that you run a software project, you accumulate technical debt—the cost of maintaining the current software and adding on patches and fixes. In a *New York Times* editorial, professors Andrew Russell and Lee Vinsel wrote that 60 percent of software development costs are spent on routine maintenance like bug fixes and upgrades.[1] Contrary to popular imagination, the enormous number of engineers and software developers that we're projected to need in the workforce in the future are not needed for new and innovative projects; 70 percent of engineers work on maintaining existing products, not making new ones.

The problem of maintenance is a good reminder that the digital world is no longer new. Like the pioneers of the first dot-com boom, it's middle-aged. If we consider the digital age to have begun with Minsky or Turing, the era is positively elderly. It's time to be more honest and realistic about what goes into technology and what it takes to keep technology working. I'm optimistic that we can find a path forward that uses technology to support both democracy and human dignity.

Mistakes were made. This could be the refrain of the media industry in dealing with the digital revolution. It could also be the refrain of the tech industry in dealing with the digital revolution. The trick is to understand the past so that we don't make the same mistakes going forward.

One thing we can do is stop referring to tech as new and shiny and innovative and instead consider it an ordinary part of life. The first computer, ENIAC, launched in 1946. We've had half a century to figure out how to integrate technology and society. That's plenty of time. Yet, after all this time, I regularly attend tech meetings in which the first ten minutes are spent awkwardly waiting while someone figures out how to use the projector to get a PowerPoint presentation to show up on the screen. Thus far, we've managed to use digital technology to increase economic inequality in the United States, facilitate illegal drug abuse, undermine the economic sustainability of the free press, cause a "fake news" crisis, roll back voting rights and fair labor protection, surveil citizens, spread junk science, harass and stalk people (primarily women and people of color) online, make flying robots that at best annoy people and at worst drop bombs, increase identity theft, enable hacks that result in millions of credit card numbers being stolen for fraudulent purposes, sell vast amounts of personal data, and elect Donald Trump to the presidency. This is not the better world that the early tech evangelists promised. It's the same world with the same types of human problems that have always existed. The problems are hidden inside code and data, which makes them harder to see and easier to ignore.

We clearly need to change our approach. We need to stop fetishizing tech. We need to audit algorithms, watch out for inequality, and reduce bias in computational systems, as well as in the tech industry. If code is law, as Lawrence Lessig writes, then we need to make sure the people who write code are doing so in accordance with the rule of law. Thus far, their efforts in self-governance have left much to be desired. We can learn from the past, but first we need to pay attention to it.

A few projects in journalism and academia suggest that a new, more balanced view of AI is on the horizon. One such project is the AI Now Institute, a policy group at NYU founded in 2017 by Kate Crawford of Microsoft Research and Meredith Whittaker of Google. Funded by Silicon Valley, the group began as a joint project of President Obama's White House Office of Science and Technology Policy and the National Economic Council. AI Now's first report focused on the near-term social and economic issues that arise from artificial intelligence technology in four fundamental areas: healthcare, labor, inequality, and ethics. Their second report issued a call to arms "for all core public institutions—such as those responsible for criminal justice, healthcare, welfare, and education—to immediately cease using 'black box' AI and algorithmic systems and to move toward systems that deliver accountability through mechanisms such as validation, auditing, or public review."[2] At Data & Society, another think tank, danah boyd is leading a group that seeks to understand and raise awareness of the role of humans in AI systems.[3] The "good selfie" experiment in chapter 9 could have benefited from a more nuanced understanding of the social context of the people making each selfie popular (and the people making the experiment). Another area worth examining is the human system for banning explicit content from social networks. Whenever violent or pornographic content is flagged online, a human must look at it and judge whether it's a video of a beheading or a photo of an object inappropriately inserted into an orifice or some other example of the worst of humanity. The psychological effects of watching streams of filth every day can be traumatic.[4] We should interrogate this practice and make a decision together, as a culture, about what it means and what should be done about it.

Inside the machine-learning community, there have been moves toward greater understanding of algorithmic inequality and accountability. The Fairness and Transparency in Machine Learning conference and community is a leader in this area.[5] Meanwhile, Harvard professor Latanya Sweeney's Data Privacy Lab at the Harvard University Institute for Quantitative Social Science is doing groundbreaking work in understanding potential violations of privacy in large data sets, especially medical data. The lab's goal is to create technology and policy "with guarantees of privacy protection while allowing society to collect and share private (or sensitive) information for many worthy purposes."[6] Also in Cambridge, the MIT Media Lab under director Joi Ito is doing admirable work to change the narrative about racial and ethnic diversity in computer science and to start interrogating systems. Prompted by MIT graduate student Karthik Dinakar's work on human-in-the-loop systems, MIT Media Lab professor Iyad Rahwan has begun working on what he calls *society-in-the-loop machine learning*, which he hopes to use to explicitly articulate moral concerns (like the trolley problem) in AI. Another project focuses on ethics and governance of AI, spearheaded by the MIT Media Lab and the Berkman Klein Center at Harvard and funded by the Ethics and Governance of Artificial Intelligence Fund.

And of course there are data journalists, who deliver at a high level despite all of the cuts to the industry. A number of remarkable tools enable sophisticated analysis of documents and data. DocumentCloud, a secure online repository for documents, contains 3.6 million source documents as of this writing and has been used by more than 8,400 journalists in 1,619 organizations worldwide. DocumentCloud is used by small and large news organizations worldwide, and it has hosted documents for high-impact stories like the Panama Papers and the Snowden documents.[7] The number of data journalists worldwide is increasing slowly. The annual NICAR conference, an annual meeting for data journalists, had over a thousand attendees for the first time in 2016. Every year, awards like the Data Journalism Awards reward truly impactful investigative-data projects. There is cause for optimism about the future.

This book has covered a lot of ground in the history and the fundamentals of today's computing technology. As I was thinking about ways people understand computers, I decided to visit the place where computing began: the Moore School of Engineering, on the University of Pennsylvania campus. There, tourists can visit the remains of the ENIAC, considered the first digital computer. In a sense, ENIAC's home was where I began too. I was born in the hospital next door to it. My parents began dating as graduate students at Penn in the 1970s. They spent a lot of time together at the campus computer center. They would arrange their punch cards into boxes, then go down to the computer center and wait their turn to put their cards through the input device to perform a statistical experiment on the mainframe computer. If you dropped the cards, my mom once told me, it was all over: good luck putting them back into the right order.

Outside the Moore building, there is a commemorative sign for the ENIAC that uses the same font as the other Philadelphia historical markers at places like Betsy Ross's house. On this day, it was crisp and clear. Dozens of high school students in dark suits and dresses streamed past me on the sidewalk. Some clutched white three-ring binders marked *Model Congress*. A few of the boys wore ski jackets over their unaccustomed suits. One had a camouflage jacket with fur around the hood. I heard a girl complaining to her friend, "I don't even know how to walk in heels," as she shuffled down the sidewalk in black suede pumps. They were cheerful, scrubbed, having a great time feeling like grownups on a college campus, without a teacher in sight. I was reminded how events like Model UN (uncomfortable clothes and all) helped my friends and me transition to adulthood. Dressing up like grownups and participating in simulated workplace activities was one of the ways we learned how to be functional adults. I suppose it would be possible to replace these kinds of experiences with videoconferences or live chats; it would be boring, however, and the data density would be low. It's hard to imagine that teenagers would want to do it.

I didn't have an access card, so I hung around the building entrance until an undergraduate with an access card showed up. He glanced at me, decided I wasn't a threat, and went back to the conversation he was muttering on his phone. I went inside and wandered, a bit lost, through the warren of hallways. I passed labs: the rapid-prototyping lab, the precision-machining lab, each filled with hulking drill presses and 3-D printers and bits of giant machinery in various states of disrepair. Engineers are hard on equipment. A mechanical-engineering student took pity on me and walked me to where ENIAC lived, in a student lounge behind wooden double doors on the first floor. Only a few panels are on display. A sign in a 1960s font proclaimed, "ENIAC, The World's First Electronic, Large- Scale, General-Purpose Digital Computer."

In the lounge, there were three diner-style wooden booths for student collaboration and three more bar areas with barstools for noncollaborating students. A stack of Facebook promotional postcards was strewn across a bar next to a computer and printer. "As a software engineer at Facebook, whether an intern or recent graduate, you will write code that impacts over 1.4 billion people around the world," a card promised. Apparently, Facebook was looking to recruit people who move fast, build things, take risks, and solve problems. "Connecting the world takes every one of us," the card assured me. I agree: connecting the world is a collaborative activity. I doubt that technology alone is the answer, however. Technology has caused the social fabric to fray in a way that suggests in-person social connections in groups and institutions are more important than ever. Understanding and group identity is best fostered through live *and* online interactions, not by interactions through screens alone.

An abandoned stack of textbooks sat next to the computer. I read their titles silently: *Life: The Science of Biology*; *Advanced Engineering Mathematics*, third edition; *Student Solutions Manual to Accompany Advanced Engineering Mathematics*, third edition. There was something nice about seeing biology and math next to each other, even though the books had been left by an absent-minded student. It boded well that the student was thinking about both natural evolution and technology.

There were three computer labs off the lounge, each filled with dozens upon dozens of computers. All of the lounge students ignored the ENIAC. Some were working on a Python problem set. Another few were discussing whether they'd studied for the MCAT yet. People wandered in with white plastic bags and fragrant foam containers: lunch from the food trucks parked outside. The students in the lab were being earnest and serious and adorable in the way of all college students. These students were what the Model UN kids wanted to grow up to be in a few years. Kids are so great. I love working at a university. Universities are such hopeful, helpful places.

ENIAC looked diminished behind its glass wall. The original took up an entire room in the basement; this small set of panels represented only a fragment of its original massive bulk. A row of vacuum tubes lay on the ground. They looked like miniature lightbulbs: the old-timey kind with glowing, exposed wires that hipster bars in Brooklyn use.

I faced the main part of the display. Black wires hung down from the bottom of each, looping from plug to plug. There were so many plugs, so many knobs. In one panel, the cycling unit, a large white eyeball, seemed to stare blankly at me. Was this the readout? I realized why it looked familiar: this is the same camera eye that Clarke and Kubrick and Minsky gave to HAL 9000, the "sentient computer" in *2001: A Space Odyssey*. The ENIAC eye is white; HAL's is red. Red is much more menacing.

Black-and-white photos on the walls showed people operating the ENIAC in its original basement home. Eight men stood arranged in front of the computer, posing stiffly. In action shots, women in suits and sensible flats with immaculately coiffed hair turned knobs and plugged things in. I've only started to see these images in the past few years. Maybe they were always around and I never noticed them. Maybe there has been a conscious attempt to put women into the visual narrative of computer science. In any case, I like it. I also liked a casual shot of the women computers, six of them, piled on top of each other laughing. I liked that these girls were having fun together. It was a reminder that computing doesn't have to be a male-dominated field. Many of the human computers of the 1940s and 1950s were women, but when the (mostly male) developers made the decision to push forward with digital computers, the women's jobs disappeared. As computing became a highly paid profession, women were also edged out. It was the result of deliberate choices. People chose to obscure the role of women in early computing and chose to exclude women from the workforce. We can change that starting now.

I thought about the distance between the ENIAC and the computers in the Windows PC and Linux PC labs. These machines represent so much human effort, so much ingenuity. I have enormous respect for the history of science and technology. What computers get wrong, though, they get wrong because they're created by humans in particular social and historical contexts. Technologists have particular disciplinary priorities that guide their decisions about developing algorithms for decision making. Often, those priorities lead them to obscure the role of human beings in creating technological systems or training data; worse, it leads them to ignore the consequences of automation on the workplace.

Looking at ENIAC, it seems absurd to imagine that this clunky chunk of metal would solve all the world's problems. Yet as ENIAC has become smaller and more powerful and we now carry it in our pockets, it's become easier to imagine things about it and project our fantasies onto it. That needs to stop. Turning real life into math is a marvelous magic trick, but too often the inconveniently human part of the equation gets pushed to the side. Humans are not now, nor have they ever been, inconvenient. Humans are the point. Humans are the beings all this technology is supposed to serve. And not just a small subset of humans, either—we should *all* be included in, and all benefit from, the development and application of technology.

Acknowledgments

Thank you to all the people who helped to bring this book to reality. I am grateful to my colleagues at New York University's (NYU's) Arthur L. Carter Journalism Institute, my colleagues at the Moore-Sloan Data Science Environment at NYU's Center for Data Science, the faculty and staff at the Tow Center for Digital Journalism at Columbia Journalism School, and my former colleagues at Temple University and the University of Pennsylvania. For reading, consulting on, or otherwise midwifing this manuscript, I am eternally indebted to Elena Lahr-Vivaz, Rosalie Siegel, Jordan Ellenberg, Cathy O'Neil, Miriam Peskowitz, Samira Baird, Lori Tharps, Kira Baker-Doyle, Jane Dmochowski, Josephine Wolff, Solon Barocas, Hanna Wallach, Katy Boss, Janet Alteveer, Leslie Hunt, Elizabeth Hunt, Kay Kinsey, Karen Masse, Stevie Santangelo, Jay Kirk, Claire Wardle, Gita Manaktala, Melinda Rankin, Kathleen Caruso, Kyle Gipson, my writers' group, and the talented team at the MIT Press. It has been an honor to participate in the community of data journalists and news nerds. I would like to thank my colleagues at the annual Computation + Journalism Symposium, everyone on the NICAR-L mailing list, and the team at ProPublica, especially Scott Klein, Derek Willis, and Celeste LeCompte. Special thanks go to Jacob Fenton, Allie Kanik, Andrew Harvard, Chase Davis, Michael Johnston, Jonathan Stray, BC Broussard, Varun D N, and everyone who assisted or advised on Bailiwick. To my family, friends, and extended family, thank you for your help and support while this book gestated. As ever, I am grateful to my husband and son for being extraordinary.

Notes

1 Hello, Reader

1. Turner, *From Counterculture to Cyberculture.*

2. Brand, "We Owe It All to the Hippies."

3. Dreyfus, *What Computers Still Can't Do.*

2 Hello, World

1. Weizenbaum, "Eliza."

2. Cerulo, *Never Saw It Coming.*

3. Miner et al., "Smartphone-Based Conversational Agents and Responses to Questions about Mental Health, Interpersonal Violence, and Physical Health."

4. Bonnington, "Tacocopter."

3 Hello, AI

1. Silver et al., "Mastering the Game of Go with Deep Neural Networks and Tree Search," 484.

2. Turing, "Computing Machinery and Intelligence."

3. Searle, "Artificial Intelligence and the Chinese Room."

4 Hello, Data Journalism

1. Cox, Bloch, and Carter, "All of Inflation's Little Parts."

2. Hart, Robbins, and Teegardin, "How the Doctors & Sex Abuse Project Came About."

3. Kestin and Maines, "Cops Hitting the Brakes—New Data Show Excessive Speeding Dropped 84% since Investigation."

4. Kunerth, "Any Way You Look at It, Florida Is the State of Weird."

5. Pierson et al., "A Large-Scale Analysis of Racial Disparities in Police Stops across the United States."

6. Angwin et al., "Machine Bias."

7. Meyer, *Precision Journalism*, 14.

8. Lewis, "Journalism in an Era of Big Data"; Diakopoulos, "Accountability in Algorithmic Decision Making"; Houston, *Computer-Assisted Reporting*; Houston and Investigative Reporters and Editors, Inc., *The Investigative Reporter's Handbook*.

9. Holovaty, "A Fundamental Way Newspaper Sites Need to Change."

10. Waite, "Announcing Politifact."

11. Holovaty, "In Memory of Chicagocrime.org."

12. Daniel and Flew, "The Guardian Reportage of the UK MP Expenses Scandal"; Flew et al., "The Promise of Computational Journalism."

13. Valentino-DeVries, Singer-Vine, and Soltani, "Websites Vary Prices, Deals Based on Users' Information."

14. Diakopoulos, "Algorithmic Accountability."

15. Anderson, "Towards a Sociology of Computational and Algorithmic Journalism"; Schudson, "Four Approaches to the Sociology of News."

16. Usher, *Interactive Journalism*.

17. Royal, "The Journalist as Programmer."

18. Hamilton, *Democracy's Detectives*.

19. Arthur, "Analysing Data Is the Future for Journalists, Says Tim Berners-Lee."

20. Silver, *The Signal and the Noise*.

5   Why Poor Schools Can't Win at StandardizedTests

1. Duncan, "Robust Data Gives Us the Roadmap to Reform."

2. Lane, "What the AP U.S. History Fight in Colorado Is Really About."

3. Broussard, "Why E-books Are Banned in My Digital Journalism Class"; Wästlund et al., "Effects of VDT and Paper Presentation on Consumption and Production of Information"; Noyes and Garland, "VDT versus Paper-Based Text"; Morineau et al.,

"The Emergence of the Contextual Role of the E-book in Cognitive Processes through an Ecological and Functional Analysis"; Noyes and Garland, "Computer- vs. Paper-Based Tasks"; Keim, "Why the Smart Reading Device of the Future May Be … Paper."

4. Ames, "Translating Magic."

5. Kraemer, Dedrick, and Sharma, "One Laptop per Child"; Purington, "One Laptop per Child."

6. Broussard, "Why Poor Schools Can't Win at Standardized Testing."

7. School District of Philadelphia, "Budget Adoption Fiscal Year 2016–2017."

6 People Problems

1. Christian and Cabell, *Initial Investigation into the Psychoacoustic Properties of Small Unmanned Aerial System Noise.*

2. Martinez, "'Drone Slayer' Claims Victory in Court."

3. Vincent, "Twitter Taught Microsoft's AI Chatbot to Be a Racist Asshole in Less than a Day."

4. Plautz, "Hitchhiking Robot Decapitated in Philadelphia."

5. Unless otherwise indicated, quotes from Minsky in this section are taken from Minsky, "Web of Stories Interview."

6. Brand, *The Media Lab*; Levy, *Hackers.*

7. Dormehl, "Why John Sculley Doesn't Wear an Apple Watch (and Regrets Booting Steve Jobs)."

8. Lewis, "Rise of the Fembots"; LaFrance, "Why Do So Many Digital Assistants Have Feminine Names?"

9. Hillis, "Radioactive Skeleton in Marvin Minsky's Closet."

10. Alba, "Chicago Uber Driver Charged with Sexual Abuse of Passenger"; Fowler, "Reflecting on One Very, Very Strange Year at Uber"; Isaac, "How Uber Deceives the Authorities Worldwide."

11. Copeland, "Summing Up Alan Turing."

12. "The Leibniz Step Reckoner and Curta Calculators—CHM Revolution."

13. Kroeger, *The Suffragents*; Shetterly, *Hidden Figures*; Grier, *When Computers Were Human.*

14. Wolfram, "Farewell, Marvin Minsky (1927–2016)."

15. Alcor Life Extension Foundation, "Official Alcor Statement Concerning Marvin Minsky."

16. Brand, "We Are As Gods."

17. Turner, *From Counterculture to Cyberculture*.

18. Brand, "We Are As Gods."

19. Hafner, *The Well*.

20. Borsook, *Cyberselfish*, 15.

21. Barlow, "A Declaration of the Independence of Cyberspace."

22. Thiel, "The Education of a Libertarian."

23. Taplin, *Move Fast and Break Things*.

24. Slovic, *The Perception of Risk*; Slovic and Slovic, *Numbers and Nerves*; Kahan et al., "Culture and Identity-Protective Cognition."

25. Leslie et al., "Expectations of Brilliance Underlie Gender Distributions across Academic Disciplines," 262.

26. Bench et al., "Gender Gaps in Overestimation of Math Performance," 158. Also see Feltman, "Men (on the Internet) Don't Believe Sexism Is a Problem in Science, Even When They See Evidence"; Williams, "The 5 Biases Pushing Women Out of STEM"; Turban, Freeman, and Waber, "A Study Used Sensors to Show That Men and Women Are Treated Differently at Work"; Moss-Racusin, Molenda, and Cramer, "Can Evidence Impact Attitudes?"; Cohoon, Wu, and Chao, "Sexism: Toxic to Women's Persistence in CSE Doctoral Programs."

27. Natanson, "A Sort of Everyday Struggle."

7   Machine Learning: The DL on ML

1. See https://xkcd.com/1425, and note that the hidden text on the online version of the comic refers to a famous anecdote about Marvin Minsky.

2. Solon, "Roomba Creator Responds to Reports of 'Poopocalypse.'"

3. Busch, "A Dozen Ways to Get Lost in Translation"; van Dalen, "The Algorithms behind the Headlines"; ACM Computing Curricula Task Force, *Computer Science Curricula 2013*.

4. IEEE Spectrum, "Tech Luminaries Address Singularity."

5. Gomes, "Facebook AI Director Yann LeCun on His Quest to Unleash Deep Learning and Make Machines Smarter."

6. "machine, *n.*"

7. Butterfield and Ngondi, *A Dictionary of Computer Science.*

8. Pedregosa et al., "Scikit-Learn: Machine Learning in Python."

9. Mitchell, "The Discipline of Machine Learning."

10. Neville-Neil, "The Chess Player Who Couldn't Pass the Salt."

11. Russell and Norvig, *Artificial Intelligence.*

12. O'Neil, *Weapons of Math Destruction.*

13. Grazian, *Mix It Up.*

14. Blow, *Fire Shut Up in My Bones.*

15. Tversky and Kahneman, "Availability." See also Kahneman, *Thinking, Fast and Slow*; Slovic, *The Perception of Risk*; Slovic and Slovic, *Numbers and Nerves*; Fischhoff and Kadvany, *Risk.*

16. See https://www.datacamp.com for more on the Titanic data science tutorial. I've omitted some parts of the tutorial for readability.

17. Quach, "Facebook Pulls Plug on Language-Inventing Chatbots?"

18. Angwin et al. "A World Apart."

19. Valentino-DeVries, Singer-Vine, and Soltani, "Websites Vary Prices, Deals Based on Users' Information."

20. Hannak et al., "Measuring Price Discrimination and Steering on E-Commerce Web Sites."

21. Heffernan, "Amazon's Prime Suspect."

22. Angwin, Mattu, and Larson, "Test Prep Is More Expensive—for Asian Students."

23. Brewster and Lynn, "Black-White Earnings Gap among Restaurant Servers."

24. Sharkey, "The Destructive Legacy of Housing Segregation."

25. Pasquale, *The Black Box Society.*

26. Lord, *A Night to Remember*; Brown, "Chronology—Sinking of S.S. TITANIC."

27. Halevy, Norvig, and Pereira, "The Unreasonable Effectiveness of Data," 8.

8   This Car Won't Drive Itself

1. "Robot Car 'Stanley' designed by Stanford Racing Team."

2. "Karel the Robot."

3. Pomerleau, "ALVINN, an Autonomous Land Vehicle in a Neural Network"; Hawkins, "Meet ALVINN, the Self-Driving Car from 1989."

4. Mundy, "Why Is Silicon Valley So Awful to Women?"

5. Oremus, "Terrifyingly Convenient."

6. DARPA Public Affairs, "Toward Machines That Improve with Experience."

7. National Highway Traffic Safety Administration and US Department of Transportation, "Federal Automated Vehicles Policy."

8. See Yoshida, "Nvidia Outpaces Intel in Robo-Car Race." Yoshida may be referring to a different standards document, in which Level 2 is equivalent to the Level 3 quoted here. Again: language and standards matter a great deal in engineering.

9. Liu et al., "CAAD: Computer Architecture for Autonomous Driving"; Thrun, "Making Cars Drive Themselves"; Thrun, "Winning the DARPA Grand Challenge."

10. See Singh, "Critical Reasons for Crashes Investigated in the National Motor Vehicle Crash Causation Survey." For more on special interest groups using statistics to construct or influence public opinion, see Best, *Damned Lies and Statistics*. Statistics are one of the ways we understand social problems, and they are often helpful in calling attention to social ills. For example, Mothers Against Drunk Driving used statistics to bring about change in drunk driving laws, which has led to greater public safety. Most people now agree that people shouldn't drive drunk. However, claiming that people shouldn't drive and machines should—that's different story.

11. Chafkin, "Udacity's Sebastian Thrun, Godfather of Free Online Education, Changes Course."

12. Marantz, "How 'Silicon Valley' Nails Silicon Valley."

13. Dougherty, "Google Photos Mistakenly Labels Black People 'Gorillas.'"

14. Evtimov et al., "Robust Physical-World Attacks on Deep Learning Models."

15. Hill, "Jamming GPS Signals Is Illegal, Dangerous, Cheap, and Easy."

16. See Harris, "God Is a Bot, and Anthony Levandowski Is His Messenger"; Marshall, "Uber Fired Its Robocar Guru, But Its Legal Fight with Google Goes On." Harris also writes that Levandowski founded a religious organization, Way of the Future, in an attempt to "develop and promote the realization of a Godhead based on Artificial Intelligence."

17. Vlasic and Boudette, "Self-Driving Tesla Was Involved in Fatal Crash, U.S. Says."

18. Tesla, Inc., "A Tragic Loss."

19. Lowy and Krisher, "Tesla Driver Killed in Crash While Using Car's 'Autopilot.'"

20. Liu et al., "CAAD: Computer Architecture for Autonomous Driving"

21. Sorrel, "Self-Driving Mercedes Will Be Programmed to Sacrifice Pedestrians to Save the Driver."

22. Taylor, "Self-Driving Mercedes-Benzes Will Prioritize Occupant Safety over Pedestrians."

23. Been, "Jaron Lanier Wants to Build a New Middle Class on Micropayments."

24. Pickrell and Li, "Driver Electronic Device Use in 2015."

25. Dadich, "Barack Obama Talks AI, Robo Cars, and the Future of the World."

9 Popular Doesn't Mean Good

1. Newman, "What Is an A-Hed?"

2. US Bureau of Labor Statistics, "Newspaper Publishers Lose over Half Their Employment from January 2001 to September 2016."

3. Pasquale, *The Black Box Society*; Gray, Bounegru, and Chambers, *The Data Journalism Handbook*; Diakopoulos, "Algorithmic Accountability"; Diakopoulos, "Accountability in Algorithmic Decision Making"; boyd and Crawford, "Critical Questions for Big Data"; Hamilton and Turner, "Accountability through Algorithm"; Cohen, Hamilton, and Turner, "Computational Journalism"; Houston, *Computer-Assisted Reporting*.

4. Angwin et al., "Machine Bias."

5. California Department of Corrections and Rehabilitation, "Fact Sheet."

6. Angwin and Larson, "Bias in Criminal Risk Scores Is Mathematically Inevitable, Researchers Say"; Kleinberg, Mullainathan, and Raghavan, "Inherent Trade-Offs in the Fair Determination of Risk Scores."

7. Bogost, "Why Nothing Works Anymore"; Brown and Duguid, *The Social Life of Information*.

8. Hempel, "Melinda Gates Has a New Mission."

9. Somerville and May, "Use of Illicit Drugs Becomes Part of Silicon Valley's Work Culture."

10. Alexander and West, *The New Jim Crow*.

11. Hern, "Silk Road Successor DarkMarket Rebrands as OpenBazaar."

12. Brown, "Nearly a Third of Millennials Have Used Venmo to Pay for Drugs."

13. Newman, "Who's Buying Drugs, Sex, and Booze on Venmo? This Site Will Tell You."

10   On the Startup Bus

1. "Jeremy Corbyn, Entrepreneur."

2. Terwiesch and Xu, "Innovation Contests, Open Innovation, and Multiagent Problem Solving."

3. Morais, "The Unfunniest Joke in Technology."

4. Tufte, *The Visual Display of Quantitative Information.*

5. Seife, *Proofiness*; Kovach and Rosenstiel, *Blur.*

11   Third-Wave AI

1. Broussard, "Artificial Intelligence for Investigative Reporting."

2. Mayer, *Dark Money*; Smith and Powell, *Dark Money, Super PACs, and the 2012 Election.*

3. Bump, "Donald Trump's Campaign Has Spent More on Hats than on Polling."

4. Donn, "Eric Trump Foundation Flouts Charity Standards."

5. Sheivachman, "Clinton vs. Trump."

6. Fletcher and Zuckerman, "Hedge Funds Battle Losses."

12   Aging Computers

1. Russell and Vinsel, "Let's Get Excited about Maintenance!"

2. Crawford, "Artificial Intelligence's White Guy Problem"; Crawford, "Artificial Intelligence—With Very Real Biases"; Campolo et al., "AI Now 2017 Report"; boyd and Crawford, "Critical Questions for Big Data."

3. boyd, Keller, and Tijerina, "Supporting Ethical Data Research"; Zook et al., "Ten Simple Rules for Responsible Big Data Research"; Elish and Hwang, "Praise the Machine! Punish the Human! The Contradictory History of Accountability in Automated Aviation."

4. Chen, "The Laborers Who Keep Dick Pics and Beheadings Out of Your Facebook Feed."

5. Fairness and Transparency in Machine Learning, "Principles for Accountable Algorithms and a Social Impact Statement for Algorithms."

6. Data Privacy Lab, "Mission Statement"; Sweeney, "Foundations of Privacy Protection from a Computer Science Perspective."

7. Pilhofer, "A Note to Users of DocumentCloud."

Bibliography

ACM Computing Curricula Task Force, ed. *Computer Science Curricula 2013: Curriculum Guidelines for Undergraduate Degree Programs in Computer Science*. New York: ACM Press, 2013. http://dl.acm.org/citation.cfm?id=2534860.

Alba, Alejandro. "Chicago Uber Driver Charged with Sexual Abuse of Passenger." *New York Daily News*, December 30, 2014. http://www.nydailynews.com/news/crime/chicago-uber-driver-charged-alleged-rape-passenger-article-1.2060817.

Alcor Life Extension Foundation. "Official Alcor Statement Concerning Marvin Minsky." Alcor News, January 27, 2016.

Alexander, Michelle, and Cornel West. *The New Jim Crow: Mass Incarceration in the Age of Colorblindness*. Revised ed. New York: New Press, 2012.

Ames, Morgan G. "Translating Magic: The Charisma of One Laptop per Child's XO Laptop in Paraguay." In *Beyond Imported Magic: Essays on Science, Technology, and Society in Latin America*, edited by Eden Medina, Ivan da Costa Marques, and Christina Holmes, 207–224. Cambridge, MA: MIT Press, 2014.

Anderson, C. W. "Towards a Sociology of Computational and Algorithmic Journalism." *New Media & Society* 15, no. 7 (November 2013): 1005–1021. doi:10.1177/1461444812465137.

Angwin, Julia, and Jeff Larson. "Bias in Criminal Risk Scores Is Mathematically Inevitable, Researchers Say." *ProPublica*, December 30, 2016. https://www.propublica.org/article/bias-in-criminal-risk-scores-is-mathematically-inevitable-researchers-say.

Angwin, Julia, Jeff Larson, Lauren Kirchner, and Surya Mattu. "A World Apart; A Joint Investigation by Consumer Reports and ProPublica Finds That Consumers in Some Minority Neighborhoods Are Charged as Much as 30 Percent More on Average for Car Insurance than in Other Neighborhoods with Similar Accident-Related Costs. What's Really Going On?" *Consumer Reports*, July 1, 2017.

Angwin, Julia, Jeff Larson, Surya Mattu, and Lauren Kirchner. "Machine Bias." *ProPublica*, May 23, 2016. https://www.propublica.org/article/machine-bias-risk-assessments-in-criminal-sentencing.

Angwin, Julia, Surya Mattu, and Jeff Larson. "Test Prep Is More Expensive—for Asian Students." *Atlantic*, September 3, 2015. https://www.theatlantic.com/educa tion/archive/2015/09/princeton-review-expensive-asian-students/403510/.

Arthur, Charles. "Analysing Data Is the Future for Journalists, Says Tim Berners-Lee." *Guardian* (US edition), November 22, 2010. https://www.theguardian.com/media/ 2010/nov/22/data-analysis-tim-berners-lee.

Barlow, John Perry. "A Declaration of the Independence of Cyberspace." Electronic Frontier Foundation, February 8, 1996. https://www.eff.org/cyberspace-indepen dence.

Been, Eric Allen. "Jaron Lanier Wants to Build a New Middle Class on Micropayments." *Nieman Lab*, May 22, 2013. http://www.niemanlab.org/2013/05/jaron -lanier-wants-to-build-a-new-middle-class-on-micropayments/.

Bench, Shane W., Heather C. Lench, Jeffrey Liew, Kathi Miner, and Sarah A. Flores. "Gender Gaps in Overestimation of Math Performance." *Sex Roles* 72, no. 11–12 (June 2015): 536–546. doi:10.1007/s11199-015-0486-9.

Best, Joel. *Damned Lies and Statistics: Untangling Numbers from the Media, Politicians, and Activists*. Updated ed. Berkeley, CA; London: University of California Press, 2012.

Blow, Charles M. *Fire Shut Up in My Bones: A Memoir*. New York: Houghton Mifflin, 2015.

Bogost, Ian. "Why Nothing Works Anymore." *Atlantic*, February 23, 2017. https:// www.theatlantic.com/technology/archive/2017/02/the-singularity-in-the-toilet -stall/517551/.

Bonnington, Christina. "Tacocopter: The Coolest Airborne Taco Delivery System That's Completely Fake." *Wired*, March 23, 2012.

Borsook, Paulina. *Cyberselfish: A Critical Romp through the Terribly Libertarian Culture of High Tech*. 1st ed. New York: PublicAffairs, 2000.

boyd, danah, and Kate Crawford. "Critical Questions for Big Data: Provocations for a Cultural, Technological, and Scholarly Phenomenon." *Information Communication and Society* 15, no. 5 (June 2012): 662–679. doi:10.1080/1369118X.2012.678878.

boyd, danah, Emily F. Keller, and Bonnie Tijerina. "Supporting Ethical Data Research: An Exploratory Study of Emerging Issues in Big Data and Technical Research." Data & Society Research Institute, August 4, 2016.

Brand, Stewart. *The Media Lab: Inventing the Future at MIT*. New York: Viking, 1987.

Brand, Stewart. "We Are As Gods." *Whole Earth Catalog* (blog), Winter 1998. http:// www.wholeearth.com/issue/1340/article/189/we.are.as.gods.

Brand, Stewart. "We Owe It All to the Hippies." *Time*, March 1, 1995. http://content
.time.com/time/magazine/article/0,9171,982602,00.html.

Brewster, Zachary W., and Michael Lynn. "Black-White Earnings Gap among Restaurant Servers: A Replication, Extension, and Exploration of Consumer Racial Discrimination in Tipping." *Sociological Inquiry* 84, no. 4 (November 2014): 545–569.
doi:10.1111/soin.12056.

Broussard, Meredith. "Artificial Intelligence for Investigative Reporting: Using an
Expert System to Enhance Journalists' Ability to Discover Original Public Affairs Stories." *Digital Journalism* 3, no. 6 (November 28, 2014): 814–831. https://doi.org/10.1
080/21670811.2014.985497.

Broussard, Meredith. "Why E-books Are Banned in My Digital Journalism Class."
New Republic, January 22, 2014. https://newrepublic.com/article/116309/data
-journalim-professor-wont-assign-e-books-heres-why.

Broussard, Meredith. "Why Poor Schools Can't Win at Standardized Testing." *Atlantic*, July 15, 2014. http://www.theatlantic.com/features/archive/2014/07/why-poor
-schools-cant-win-at-standardized-testing/374287/.

Brown, David G. "Chronology—Sinking of S.S. TITANIC." *Encyclopedia Titanica*. Last
updated June 9, 2009. https://www.encyclopedia-titanica.org/articles/et_timeline
.pdf.

Brown, John Seely, and Paul Duguid. *The Social Life of Information*. Updated, with a
new preface. Boston: Harvard Business Review Press, 2017.

Brown, Mike. "Nearly a Third of Millennials Have Used Venmo to Pay for
Drugs." *LendEDU.com* (blog), July 10, 2017. https://lendedu.com/blog/nearly-third
-millennials-used-venmo-pay-drugs/.

Bump, Philip. "Donald Trump's Campaign Has Spent More on Hats than on Polling." *The Washington Post*, October 25, 2016. https://www.washingtonpost.com/
news/the-fix/wp/2016/10/25/donald-trumps-campaign-has-spent-more-on-hats
-than-on-polling.

Busch, Lawrence. "A Dozen Ways to Get Lost in Translation: Inherent Challenges in
Large-Scale Data Sets." *International Journal of Communication* 8 (2014): 1727–1744.

Butterfield, A., and Gerard Ekembe Ngondi, eds. *A Dictionary of Computer Science*. 7th
ed. Oxford Quick Reference. Oxford, UK; New York: Oxford University Press, 2016.

California Department of Corrections and Rehabilitation. "Fact Sheet: COMPAS
Assessment Tool Launched—Evidence-Based Rehabilitation for Offender Success,"
April 15, 2009. http://www.cdcr.ca.gov/rehabilitation/docs/FS_COMPAS_Final_4
-15-09.pdf.

Campolo, Alex, Madelyn Sanfilippo, Meredith Whittaker, Kate Crawford, Andrew Selbst, and Solon Barocas. "AI Now 2017 Report." AI Now Institute, New York University, October 18, 2017. https://assets.contentful.com/8wprhhvnpfc0/1A9c3ZTCZ a2KEYM64Wsc2a/8636557c5fb14f2b74b2be64c3ce0c78/_AI_Now_Institute_2017 _Report_.pdf.

Cerulo, Karen A. *Never Saw It Coming: Cultural Challenges to Envisioning the Worst.* Chicago: University of Chicago Press, 2006.

Chafkin, Max. "Udacity's Sebastian Thrun, Godfather of Free Online Education, Changes Course." *Fast Company,* November 14, 2013. https://www.fastcompany .com/3021473/udacity-sebastian-thrun-uphill-climb.

Chen, Adrian. "The Laborers Who Keep Dick Pics and Beheadings Out of Your Facebook Feed." *Wired,* October 23, 2014. https://www.wired.com/2014/10/content -moderation/.

Christian, Andrew, and Randolph Cabell. *Initial Investigation into the Psychoacoustic Properties of Small Unmanned Aerial System Noise.* Hampton, VA: NASA Langley Research Center, American Institute of Aeronautics and Astronautics, 2017. https:// ntrs.nasa.gov/archive/nasa/casi.ntrs.nasa.gov/20170005870.pdf.

Cohen, Sarah, James T. Hamilton, and Fred Turner. "Computational Journalism." *Communications of the ACM* 54, no. 10 (October 1, 2011): 66. doi:10.1145/2001269 .2001288.

Cohoon, J. McGrath, Zhen Wu, and Jie Chao. "Sexism: Toxic to Women's Persistence in CSE Doctoral Programs," 158. New York: ACM Press, 2009. https://doi .org/10.1145/1508865.1508924.

Copeland, Jack. "Summing Up Alan Turing." *Oxford University Press* (blog), November 29, 2012. https://blog.oup.com/2012/11/summing-up-alan-turing/.

Cox, Amanda, Matthew Bloch, and Shan Carter. "All of Inflation's Little Parts." *New York Times,* May 3, 2008. http://www.nytimes.com/interactive/2008/05/03/business/ 20080403_SPENDING_GRAPHIC.html.

Crawford, Kate. "Artificial Intelligence—With Very Real Biases." *Wall Street Journal,* October 17, 2017. https://www.wsj.com/articles/artificial-intelligencewith-very-real -biases-1508252717.

Crawford, Kate. "Artificial Intelligence's White Guy Problem." *New York Times,* June 26, 2016. https://www.nytimes.com/2016/06/26/opinion/sunday/artificial -intelligences-white-guy-problem.html.

Dadich, Scott. "Barack Obama Talks AI, Robo Cars, and the Future of the World." *Wired,* November 2016. https://www.wired.com/2016/10/president-obama-mit-joi -ito-interview/.

Daniel, Anna, and Terry Flew. "The Guardian Reportage of the UK MP Expenses Scandal: A Case Study of Computational Journalism." In *Communications Policy and Research Forum 2010*, November 15–16, 2010. https://www.researchgate.net/publication/279424256_The_Guardian_Reportage_of_the_UK_MP_Expenses_Scandal_A_Case_Study_of_Computational_Journalism.

DARPA Public Affairs. "Toward Machines That Improve with Experience," March 16, 2017. https://www.darpa.mil/news-events/2017-03-16.

Data Privacy Lab. "Mission Statement," n.d. https://dataprivacylab.org/about.html.

Diakopoulos, Nicholas. "Accountability in Algorithmic Decision Making." *Communications of the ACM* 59, no. 2 (January 25, 2016): 56–62. doi:10.1145/2844110.

Diakopoulos, Nicholas. "Algorithmic Accountability: Journalistic Investigation of Computational Power Structures." *Digital Journalism* 3, no. 3 (November 7, 2014) 398–415. https://doi.org/10.1080/21670811.2014.976411.

Donn, Jeff. "Eric Trump Foundation Flouts Charity Standards." *AP News*, December 23, 2016. https://apnews.com/760b4159000b4a1cb1901cb038021cea.

Dormehl, Luke. "Why John Sculley Doesn't Wear an Apple Watch (and Regrets Booting Steve Jobs)." *Cult of Mac*, February 19, 2016. https://www.cultofmac.com/413044/john-sculley-apple-watch-steve-jobs/.

Dougherty, Conor. "Google Photos Mistakenly Labels Black People 'Gorillas.'" *New York Times*, July 1, 2015. https://bits.blogs.nytimes.com/2015/07/01/google-photos-mistakenly-labels-black-people-gorillas/.

Dreyfus, Hubert L. *What Computers Still Can't Do: A Critique of Artificial Reason*. Cambridge, MA: MIT Press, 1992.

Duncan, Arne. "Robust Data Gives Us the Roadmap to Reform." Paper presented at the Fourth Annual IES Research Conference, June 8, 2009. https://www.ed.gov/news/speeches/robust-data-gives-us-roadmap-reform.

Elish, Madeleine, and Tim Hwang. "Praise the Machine! Punish the Human! The Contradictory History of Accountability in Automated Aviation." Comparative Studies in Intelligent Systems—Working Paper #1. Intelligence and Autonomy Initiative: Data & Society Research Institute, February 24, 2015. https://datasociety.net/pubs/ia/Elish-Hwang_AccountabilityAutomatedAviation.pdf.

Evtimov, Ivan, Kevin Eykholt, Earlence Fernandes, Tadayoshi Kohno, Bo Li, Atul Prakash, Amir Rahmati, and Dawn Song. "Robust Physical-World Attacks on Deep Learning Models." In *arXiv Preprint 1707.08945*, 2017.

Fairness and Transparency in Machine Learning. "Principles for Accountable Algorithms and a Social Impact Statement for Algorithms," n.d. https://www.fatml.org/resources/principles-for-accountable-algorithms.

Feltman, Rachel. "Men (on the Internet) Don't Believe Sexism Is a Problem in Science, Even When They See Evidence," January 8, 2015.

Fischhoff, Baruch, and John Kadvany. *Risk: A Very Short Introduction*. Oxford: Oxford University Press, 2011.

Fletcher, Laurence, and Gregory Zuckerman. "Hedge Funds Battle Losses," 2016. http://ezproxy.library.nyu.edu:2048/login?url=http://search.proquest.com/docview /1811735200?accountid=12768.

Flew, Terry, Christina Spurgeon, Anna Daniel, and Adam Swift. "The Promise of Computational Journalism." *Journalism Practice* 6, no. 2 (April 2012): 157–171. doi:1 0.1080/17512786.2011.616655.

Fowler, Susan J. "Reflecting on One Very, Very Strange Year at Uber." *Susan Fowler* (blog), February 19, 2017. https://www.susanjfowler.com/blog/2017/2/19/reflecting -on-one-very-strange-year-at-uber.

Gomes, Lee. "Facebook AI Director Yann LeCun on His Quest to Unleash Deep Learning and Make Machines Smarter." *IEEE Spectrum* (blog), February 18, 2015. http://spectrum.ieee.org/automaton/robotics/artificial-intelligence/facebook-ai -director-yann-lecun-on-deep-learning.

Gray, Jonathan, Liliana Bounegru, and Lucy Chambers, eds. *The Data Journalism Handbook: How Journalists Can Use Data to Improve the News*. Sebastopol, CA: O'Reilly Media, 2012.

Grazian, David. *Mix It Up: Popular Culture, Mass Media, and Society*. 2nd ed. New York: W. W. Norton, 2017.

Grier, David Alan. *When Computers Were Human*. Princeton, NJ: Princeton University Press, 2007.

Hafner, Katie. *The Well: A Story of Love, Death, and Real Life in the Seminal Online Community*. New York: Carroll & Graf, 2001.

Halevy, Alon, Peter Norvig, and Fernando Pereira. "The Unreasonable Effectiveness of Data." *IEEE Intelligent Systems* 24, no. 2 (March 2009): 8–12. https://doi.org/ 10.1109/MIS.2009.36.

Hamilton, James. 2016. *Democracy's Detectives: The Economics of Investigative Journalism*. Cambridge, MA: Harvard University Press.

Hamilton, James T., and Fred Turner. "Accountability through Algorithm: Developing the Field of Computational Journalism." Paper presented at the Center for Advanced Study in the Behavioral Sciences Summer Workshop, July 2009. http:// web.stanford.edu/~fturner/Hamilton%20Turner%20Acc%20by%20Alg%20Final .pdf.

Hannak, Aniko, Gary Soeller, David Lazer, Alan Mislove, and Christo Wilson. "Measuring Price Discrimination and Steering on E-Commerce Web Sites." In *Proceedings of the 2014 Internet Measurement Conference*, 305–318. New York: ACM Press, 2014. doi:10.1145/2663716.2663744.

Harris, Mark. "God Is a Bot, and Anthony Levandowski Is His Messenger." *Wired*, September 27, 2017. https://www.wired.com/story/god-is-a-bot-and-anthony -levandowski-is-his-messenger/.

Hart, Ariel, Danny Robbins, and Carrie Teegardin. "How the Doctors & Sex Abuse Project Came About." *Atlanta Journal-Constitution*, July 6, 2016. http://doctors.ajc. com/about_this_investigation/.

Hawkins, Andrew J. "Meet ALVINN, the Self-Driving Car from 1989." *The Verge*, November 27, 2016. http://www.theverge.com/2016/11/27/13752344/alvinn-self -driving-car-1989-cmu-navlab.

Heffernan, Virginia. "Amazon's Prime Suspect." *New York Times*, August 6, 2010. http://www.nytimes.com/2010/08/08/magazine/08FOB-medium-t.html.

Hempel, Jessi. "Melinda Gates Has a New Mission: Women in Tech." *Wired*, Backchannel, September 28, 2016. https://backchannel.com/melinda-gates-has-a-new -mission-women-in-tech-8eb706d0a903.

Hern, Alex. "Silk Road Successor DarkMarket Rebrands as OpenBazaar." *The Guardian*, April 30, 2014. https://www.theguardian.com/technology/2014/apr/30/silk -road-darkmarket-openbazaar-online-drugs-marketplace.

Hill, Kashmir. "Jamming GPS Signals Is Illegal, Dangerous, Cheap, and Easy." *Gizmodo*, July 24, 2017. https://gizmodo.com/jamming-gps-signals-is-illegal-dangerous -cheap-and-e-1796778955.

Hillis, W. Daniel. "Radioactive Skeleton in Marvin Minsky's Closet." Paper presented at the Web of Stories, n.d. https://webofstories.com/play/danny.hillis/174.

Holovaty, Adrian. "A Fundamental Way Newspaper Sites Need to Change." Holovaty.com, September 6, 2006. http://www.holovaty.com/writing/fundamental -change/.

Holovaty, Adrian. "In Memory of Chicagocrime.org." Holovaty.com, January 31, 2008. http://www.holovaty.com/writing/chicagocrime.org-tribute/.

Houston, Brant. *Computer-Assisted Reporting: A Practical Guide*. 4th ed. New York: Routledge, 2015.

Houston, Brant, and Investigative Reporters and Editors, Inc., eds. *The Investigative Reporter's Handbook: A Guide to Documents, Databases, and Techniques*. 5th ed. Boston: Bedford/St. Martin's, 2009.

IEEE Spectrum. "Tech Luminaries Address Singularity." *IEEE Spectrum*, June 1, 2008. http://spectrum.ieee.org/computing/hardware/tech-luminaries-address-singularity.

Isaac, Mike. "How Uber Deceives the Authorities Worldwide." *New York Times*, March 3, 2017. https://www.nytimes.com/2017/03/03/technology/uber-greyball-program-evade-authorities.html.

"Jeremy Corbyn, Entrepreneur." *Economist*, June 15, 2017. http://www.economist.com/news/britain/21723426-labours-leader-has-disrupted-business-politics-jeremy-corbyn-entrepreneur.

Kahan, Dan M., Donald Braman, John Gastil, Paul Slovic, and C. K. Mertz. "Culture and Identity-Protective Cognition: Explaining the White-Male Effect in Risk Perception." *Journal of Empirical Legal Studies* 4, no. 3 (November 2007): 465–505.

Kahneman, Daniel. *Thinking, Fast and Slow*. New York: Farrar, Straus and Giroux, 2013.

"Karel the Robot: Fundamentals." Middle Tennessee State University, n.d. Accessed April 14, 2017. https://cs.mtsu.edu/~untch/karel/fundamentals.html.

Keim, Brandon. "Why the Smart Reading Device of the Future May Be … Paper." *Wired*, May 1, 2014. https://www.wired.com/2014/05/reading-on-screen-versus-paper/.

Kestin, Sally, and John Maines. "Cops Hitting the Brakes—New Data Show Excessive Speeding Dropped 84% since Investigation." *Sun Sentinel* (Fort Lauderdale, FL), December 30, 2012.

Kleinberg, J., S. Mullainathan, and M. Raghavan. "Inherent Trade-Offs in the Fair Determination of Risk Scores." *ArXiv E-Prints*, September 2016.

Kovach, Bill, and Tom Rosenstiel. *Blur: How to Know What's True in the Age of Information Overload*. New York: Bloomsbury, 2011.

Kraemer, Kenneth L., Jason Dedrick, and Prakul Sharma. "One Laptop per Child: Vision vs. Reality." *Communications of the ACM* 52, no. 6 (June 1, 2009): 66. doi:10.1145/1516046.1516063.

Kroeger, Brooke. 2017. *The Suffragents: How Women Used Men to Get the Vote*. Albany: State University of New York Press.

Kunerth, Jeff. "Any Way You Look at It, Florida Is the State of Weird." *Orlando Sentinel*, June 13, 2013. http://articles.orlandosentinel.com/2013-06-13/features/os-florida-is-weird-20130613_1_florida-state-weird-florida-central-florida-lakes.

LaFrance, Adrienne. 2016. Why Do So Many Digital Assistants Have Feminine Names? *Atlantic*, March 30, 2016. https://www.theatlantic.com/technology/archive/2016/03/why-do-so-many-digital-assistants-have-feminine-names/475884/.

Lane, Charles. "What the AP U.S. History Fight in Colorado Is Really About." *Washington Post*, November 6, 2014. https://www.washingtonpost.com/blogs/post -partisan/wp/2014/11/06/what-the-ap-u-s-history-fight-in-colorado-is-really-about.

"The Leibniz Step Reckoner and Curta Calculators." Computer History Museum, n.d. Accessed April 14, 2017. http://www.computerhistory.org/revolution/calculators/ 1/49.

Leslie, S.-J., A. Cimpian, M. Meyer, and E. Freeland. "Expectations of Brilliance Underlie Gender Distributions across Academic Disciplines." *Science* 347, no. 6219 (January 16, 2015): 262–265. doi:10.1126/science.1261375.

Lewis, Seth C. "Journalism in an Era of Big Data: Cases, Concepts, and Critiques." *Digital Journalism* 3, no. 3 (November 27, 2014): 321–330. https://doi.org/10.1080/ 21670811.2014.976399.

Lewis, Tanya. "Rise of the Fembots: Why Artificial Intelligence Is Often Female." *LiveScience*, February 19, 2015.

Levy, Steven. 2010. *Hackers*. 1st ed. Sebastopol, CA: O'Reilly Media.

Liu, Shaoshan, Jie Tang, Zhe Zhang, and Jean-Luc Gaudiot. "CAAD: Computer Architecture for Autonomous Driving." *CoRR* abs/1702.01894 (February 7, 2017). http://arxiv.org/abs/1702.01894.

Lord, Walter. *A Night to Remember*. New York: Henry, Holt, and Co., 2005.

Lowy, Joan, and Tom Krisher. "Tesla Driver Killed in Crash While Using Car's 'Auto-pilot.'" *Associated Press*, June 30, 2016. http://www.bigstory.ap.org/article/ee71bd075 fb948308727b4bbff7b3ad8/self-driving-car-driver-died-after-crash-florida-first.

"machine, *n.*" *OED Online*, Oxford University Press. Last updated March 2000. http://www.oed.com/view/Entry/111850.

Marantz, Andrew. "How 'Silicon Valley' Nails Silicon Valley." *New Yorker*, June 9, 2016. http://www.newyorker.com/culture/culture-desk/how-silicon-valley -nails-silicon-valley.

Marshall, Aarian. "Uber Fired Its Robocar Guru, But Its Legal Fight with Google Goes On." *Wired*, May 30, 2017. https://www.wired.com/2017/05/uber-fires-anthony -levandowski-waymo-google-lawsuit/.

Martinez, Natalia. "'Drone Slayer' Claims Victory in Court." *WAVE 3 News*, October 26, 2015. http://www.wave3.com/story/30355558/drone-slayer-claims-victory-in -court.

Mayer, Jane. *Dark Money: The Hidden History of the Billionaires behind the Rise of the Radical Right*. New York: Doubleday, 2016.

Meyer, Philip. *Precision Journalism: A Reporter's Introduction to Social Science Methods.* 4th ed. Lanham, MD: Rowman & Littlefield Publishers, 2002.

Miner, Adam S., Arnold Milstein, Stephen Schueller, Roshini Hegde, Christina Mangurian, and Eleni Linos. "Smartphone-Based Conversational Agents and Responses to Questions about Mental Health, Interpersonal Violence, and Physical Health." *JAMA Internal Medicine* 176, no. 5 (May 1, 2016): 619. doi:10.1001/jamainternmed .2016.0400.

Minsky, Marvin. "Web of Stories Interview: Marvin Minsky." Web of Stories, January 29, 2011. https://www.webofstories.com/play/marvin.minsky/1.

Mitchell, Tom M. "The Discipline of Machine Learning." Pittsburgh, PA: School of Computer Science, Carnegie Mellon University, July 2006. http://reports-archive .adm.cs.cmu.edu/anon/ml/abstracts/06-108.html.

Morais, Betsy. "The Unfunniest Joke in Technology." *New Yorker*, September 9, 2013. https://www.newyorker.com/tech/elements/the-unfunniest-joke-in-technology.

Morineau, Thierry, Caroline Blanche, Laurence Tobin, and Nicolas Guéguen. "The Emergence of the Contextual Role of the E-book in Cognitive Processes through an Ecological and Functional Analysis." *International Journal of Human-Computer Studies* 62, no. 3 (March 2005): 329–348. doi:10.1016/j.ijhcs.2004.10.002.

Moss-Racusin, Corinne A., Aneta K. Molenda, and Charlotte R. Cramer. "Can Evidence Impact Attitudes? Public Reactions to Evidence of Gender Bias in STEM Fields." *Psychology of Women Quarterly* 39 (2) (June 2015): 194–209. https://doi .org/10.1177/0361684314565777.

Mundy, Liza. "Why Is Silicon Valley So Awful to Women?" *The Atlantic*, April 2017. https://www.theatlantic.com/magazine/archive/2017/04/why-is-silicon-valley-so -awful-to-women/517788/.

Natanson, Hannah. "A Sort of Everyday Struggle." *The Harvard Crimson*, October 20, 2017. https://www.thecrimson.com/article/2017/10/20/everyday-struggle-women -math/.

National Highway Traffic Safety Administration and US Department of Transportation. "Federal Automated Vehicles Policy: Accelerating the Next Revolution in Roadway Safety," September 2016.

Neville-Neil, George V. "The Chess Player Who Couldn't Pass the Salt." *Communications of the ACM* 60, no. 4 (March 24, 2017): 24–25. doi:10.1145/3055277.

Newman, Barry. "What Is an A-Hed?" *Wall Street Journal*, November 15, 2010. https://www.wsj.com/articles/SB10001424052702303362404575580494180594982.

Newman, Lily Hay. "Who's Buying Drugs, Sex, and Booze on Venmo? This Site Will Tell You." *Future Tense: The Citizen's Guide to the Future*, February 23, 2015. http://

www.slate.com/blogs/future_tense/2015/02/23/vicemo_shows_venmo_transac
tions_related_to_drugs_alcohol_and_sex.html.

Noyes, Jan M., and Kate J. Garland. "VDT versus Paper-Based Text: Reply to Mayes, Sims and Koonce." *International Journal of Industrial Ergonomics* 31, no. 6 (June 2003): 411–423. doi:10.1016/S0169-8141(03)00027-1.

Noyes, Jan M., and Kate J. Garland. "Computer- vs. Paper-Based Tasks: Are They Equivalent?" *Ergonomics* 51, no. 9 (September 2008): 1352–1375. doi:10.1080/ 00140130802170387.

O'Neil, Cathy. *Weapons of Math Destruction: How Big Data Increases Inequality and Threatens Democracy.* 1st ed. New York: Crown Publishers, 2016.

Oremus, Will. "Terrifyingly Convenient." *Slate*, April 3, 2016. http://www.slate. com/articles/technology/cover_story/2016/04/alexa_cortana_and_siri_aren_t_novel ties_anymore_they_re_our_terrifyingly.html.

Pasquale, Frank. *The Black Box Society: The Secret Algorithms That Control Money and Information.* Cambridge, MA: Harvard University Press, 2015.

Pedregosa, F., G. Varoquaux, A. Gramfort, V. Michel, B. Thirion, O. Grisel, M. Blondel, et al. "Scikit-Learn: Machine Learning in Python." *Journal of Machine Learning Research* 12 (2011): 2825–2830.

Pickrell, Timothy M., and Hongying (Ruby) Li. "Driver Electronic Device Use in 2015." Traffic Safety Facts Research Note. Washington, DC: National Highway Traffic Safety Administration, September 2016. https://www.nhtsa.gov/sites/nhtsa.dot .gov/files/documents/driver_electronic_device_use_in_2015_0.pdf.

Pierson, E., C. Simoiu, J. Overgoor, S. Corbett-Davies, V. Ramachandran, C. Phillips, and S. Goel. "A Large-Scale Analysis of Racial Disparities in Police Stops across the United States." Stanford Open Policing Project. Stanford University, 2017. https://5harad.com/papers/traffic-stops.pdf.

Pilhofer, Aron. "A Note to Users of DocumentCloud." Medium, July 27, 2017. https://medium.com/@pilhofer/a-note-to-users-of-documentcloud-org-264177466 1bb.

Plautz, Jessica. "Hitchhiking Robot Decapitated in Philadelphia." *Mashable*, August 1, 2015. http://mashable.com/2015/08/01/hitchbot-destroyed.

Pomerleau, Dean A. "ALVINN, an Autonomous Land Vehicle in a Neural Network." Carnegie Mellon University, 1989. http://repository.cmu.edu/cgi/viewcontent.cgi?ar ticle=2874&context=compsci.

Purington, David. "One Laptop per Child: A Misdirection of Humanitarian Effort." *ACM SIGCAS Computers and Society* 40, no. 1 (March 1, 2010): 28–33. doi:10.1145/ 1750888.1750892.

Quach, Katyanna. "Facebook Pulls Plug on Language-Inventing Chatbots? The Truth." *Register*, August 1, 2017. https://www.theregister.co.uk/2017/08/01/facebook _chatbots_did_not_invent_new_language.

"Robot Car 'Stanley' Designed by Stanford Racing Team." Stanford Racing Team, 2005. http://cs.stanford.edu/group/roadrunner/stanley.html.

Royal, Cindy. "The Journalist as Programmer: A Case Study of the *New York Times* Interactive News Technology Department." University of Texas at Austin, April 2010. http://www.cindyroyal.com/present/royal_isoj10.pdf.

Russell, Andrew, and Lee Vinsel. "Let's Get Excited about Maintenance!" *New York Times*, July 22, 2017. https://mobile.nytimes.com/2017/07/22/opinion/sunday/lets -get-excited-about-maintenance.html.

Russell, Stuart J., and Peter Norvig. *Artificial Intelligence: A Modern Approach*. 3rd ed. Harlow, UK: Pearson, 2016.

School District of Philadelphia. "Budget Adoption Fiscal Year 2016–2017." May 26, 2016. http://webgui.phila.k12.pa.us/uploads/jq/BX/jqBX-vKcX2GM7Nbrpgqwzg/ FY17-Budget-Adoption_FINAL_5.26.16.pdf.

Schudson, Michael. "Four Approaches to the Sociology of News." In *Mass Media and Society*, 4th ed., edited by James Curran and Michael Gurevitch, 172–197. London: Hodder Arnold, 2005.

"Scientists Propose a Novel Regional Path Tracking Scheme for Autonomous Ground Vehicles." Phys Org, January 16, 2017. https://phys.org/news/2017-01-scientists- regional-path-tracking-scheme.html.

Searle, John R. "Artificial Intelligence and the Chinese Room: An Exchange." *New York Review of Books*, February 16, 1989. http://www.nybooks.com/articles/1989/ 02/16/artificial-intelligence-and-the-chinese-room-an-ex/.

Seife, Charles. *Proofiness: How You're Being Fooled by the Numbers*. New York: Penguin, 2011.

Sharkey, Patrick. "The Destructive Legacy of Housing Segregation." *Atlantic*, June 2016. https://www.theatlantic.com/magazine/archive/2016/06/the-eviction-curse/ 480738/.

Sheivachman, Andrew. "Clinton vs. Trump: Where Presidential Candidates Spend Their Travel Dollars." *Skift*, October 4, 2016. https://skift.com/2016/10/04/clinton -vs-trump-where-presidential-candidates-spend-their-travel-dollars/.

Shetterly, Margot Lee. *Hidden Figures: The American Dream and the Untold Story of the Black Women Mathematicians Who Helped Win the Space Race*. New York: HarperCollins, 2016.

Silver, David, Aja Huang, Chris J. Maddison, Arthur Guez, Laurent Sifre, George van den Driessche, Julian Schrittwieser, et al. "Mastering the Game of Go with Deep Neural Networks and Tree Search." *Nature* 529 (January 28, 2016): 484–489. doi:10.1038/nature16961.

Silver, Nate. *The Signal and the Noise: Why so Many Predictions Fail—but Some Don't.* New York: Penguin Books, 2015.

Singh, Santokh. "Critical Reasons for Crashes Investigated in the National Motor Vehicle Crash Causation Survey." Traffic Safety Facts Crash Stats. Washington, DC: Bowhead Systems Management, Inc., working under contract with the Mathematical Analysis Division of the National Center for Statistics and Analysis, NHTSA, February 2015. https://crashstats.nhtsa.dot.gov/Api/Public/ViewPublication/812115.

Slovic, Paul. *The Perception of Risk.* Risk, Society, and Policy Series. Sterling, VA: Earthscan Publications, 2000.

Slovic, S., and P. Slovic, eds. *Numbers and Nerves: Information, Emotion, and Meaning in a World of Data.* Corvallis: Oregon State University Press, 2015.

Smith, Melissa M., and Larry Powell. *Dark Money, Super PACs, and the 2012 Election. Lexington Studies in Political Communication.* Lanham, MD: Lexington Books, 2013.

Solon, Olivia. "Roomba Creator Responds to Reports of 'Poopocalypse': 'We See This a Lot.'" *Guardian* (US edition), August 15, 2016. https://www.theguardian.com/technology/2016/aug/15/roomba-robot-vacuum-poopocalypse-facebook-post.

Somerville, Heather, and Patrick May. "Use of Illicit Drugs Becomes Part of Silicon Valley's Work Culture." *San Jose Mercury News,* July 25, 2014. http://www.mercurynews.com/2014/07/25/use-of-illicit-drugs-becomes-part-of-silicon-valleys-work-culture/.

Sorrel, Charlie. "Self-Driving Mercedes Will Be Programmed to Sacrifice Pedestrians to Save the Driver." October 13, 2016. https://www.fastcompany.com/3064539/self-driving-mercedes-will-be-programmed-to-sacrifice-pedestrians-to-save-the-driver.

Sweeney, Latanya. "Foundations of Privacy Protection from a Computer Science Perspective." Carnegie Mellon University, 2000. http://repository.cmu.edu/isr/245/.

Taplin, Jonathan. *Move Fast and Break Things: How Facebook, Google, and Amazon Cornered Culture and Undermined Democracy.* New York: Little, Brown and Co., 2017.

Taylor, Michael. "Self-Driving Mercedes-Benzes Will Prioritize Occupant Safety over Pedestrians." *Car and Driver* (blog), October 7, 2016. https://blog.caranddriver.com/self-driving-mercedes-will-prioritize-occupant-safety-over-pedestrians/.

Terwiesch, Christian, and Yi Xu. "Innovation Contests, Open Innovation, and Multiagent Problem Solving." *Management Science* 54, no. 9 (September 2008): 1529–1543. doi:10.1287/mnsc.1080.0884.

Tesla, Inc. "A Tragic Loss," June 30, 2016. https://www.tesla.com/blog/tragic-loss.

Thiel, Peter. "The Education of a Libertarian." *Cato Unbound* (blog), April 13, 2009. https://www.cato-unbound.org/2009/04/13/peter-thiel/education-libertarian.

Thrun, S. "Winning the DARPA Grand Challenge: A Robot Race through the Mojave Desert," 11. IEEE, 2006. https://doi.org/10.1109/ASE.2006.74.

Thrun, Sebastian. "Making Cars Drive Themselves," 1–86. IEEE, 2008. https://doi .org/10.1109/HOTCHIPS.2008.7476533.

Tufte, Edward R. 2001. *The Visual Display of Quantitative Information*. 2nd ed. Cheshire, CT: Graphics Press.

Turban, Stephen, Laura Freeman, and Ben Waber. "A Study Used Sensors to Show That Men and Women Are Treated Differently at Work." *Harvard Business Review*, October 23, 2017. https://hbr.org/2017/10/a-study-used-sensors-to-show-that-men -and-women-are-treated-differently-at-work.

Turing, A. M. "Computing Machinery and Intelligence." *Mind* 59, no. 236 (1950): 433–460.

Turner, Fred. *From Counterculture to Cyberculture: Stewart Brand, the Whole Earth Network, and the Rise of Digital Utopianism*. Chicago: University of Chicago Press, 2008.

Tversky, Amos, and Daniel Kahneman. "Availability: A Heuristic for Judging Frequency and Probability." *Cognitive Psychology* 5, no. 2 (September 1973): 207–232. doi:10.1016/0010-0285(73)90033-9.

US Bureau of Labor Statistics. "Newspaper Publishers Lose over Half Their Employment from January 2001 to September 2016." TED: The Economics Daily, April 3, 2017. https://www.bls.gov/opub/ted/2017/newspaper-publishers-lose-over -half-their-employment-from-january-2001-to-september-2016.htm.

Usher, Nikki. *Interactive Journalism: Hackers, Data, and Code*. Urbana: University of Illinois Press, 2016.

Valentino-DeVries, Jennifer, Jeremy Singer-Vine, and Ashkan Soltani. "Websites Vary Prices, Deals Based on Users' Information." *Wall Street Journal*, December 24, 2012. https://www.wsj.com/articles/SB10001424127887323777204578189391813881534.

van Dalen, Arjen. "The Algorithms behind the Headlines: How Machine-Written News Redefines the Core Skills of Human Journalists." *Journalism Practice* 6, no. 5–6 (October 2012): 648–658. doi:10.1080/17512786.2012.667268.

Vincent, James. "Twitter Taught Microsoft's AI Chatbot to Be a Racist Asshole in Less than a Day." *The Verge*, March 24, 2016. https://www.theverge.com/2016/3/24/11297050/tay-microsoft-chatbot-racist.

Vlasic, Bill, and Neal E. Boudette. "Self-Driving Tesla Was Involved in Fatal Crash, U.S. Says." *New York Times*, June 30, 2016. https://www.nytimes.com/2016/07/01/business/self-driving-tesla-fatal-crash-investigation.html.

Waite, Matt. "Announcing Politifact." *MattWaite.com* (blog), August 22, 2007. http://www.mattwaite.com/posts/2007/aug/22/announcing-politifact/.

Wästlund, Erik, Henrik Reinikka, Torsten Norlander, and Trevor Archer. "Effects of VDT and Paper Presentation on Consumption and Production of Information: Psychological and Physiological Factors." *Computers in Human Behavior* 21, no. 2 (March 2005): 377–394. doi:10.1016/j.chb.2004.02.007.

Weizenbaum, Joseph. "Eliza," n.d. http://www.atariarchives.org/bigcomputergames/showpage.php?page=23.

Williams, Joan C. "The 5 Biases Pushing Women Out of STEM." *Harvard Business Review*, March 24, 2015. https://hbr.org/2015/03/the-5-biases-pushing-women-out-of-stem.

Wolfram, Stephen. "Farewell, Marvin Minsky (1927–2016)." *Stephen Wolfram* (blog), January 26, 2016. http://blog.stephenwolfram.com/2016/01/farewell-marvin-minsky-19272016/.

Yoshida, Junko. "Nvidia Outpaces Intel in Robo-Car Race." *EE Times*, October 11, 2017. https://www.eetimes.com/document.asp?doc_id=1332425.

Zook, Matthew, Solon Barocas, danah boyd, Kate Crawford, Emily Keller, Seeta Peña Gangadharan, Alyssa Goodman, et al. "Ten Simple Rules for Responsible Big Data Research." Edited by Fran Lewitter. *PLOS Computational Biology* 13, no. 3 (March 30, 2017): e1005399. https://doi.org/10.1371/journal.pcbi.1005399.

Index

Abacus, 75
Ability beliefs, 83
Academy at Palumbo, 56–57
Ackerman, Arlene, 58–59
Activism, cyberspace, 82–83
Adair, Bill, 45
AI Now Institute, 194–195
AirBnB, 168
Albrecht, Steve, 159
Alda, Alan, 70
Alexa, 38–39, 72
Alexander, Michelle, 159
Algorithmic accountability reporting, 7,
    43–44, 65–66
Algorithms
  bias in, 44, 150, 155–157, 195
  defined, 7, 94
  elevator, 157
  function of, 43–44
  risk, 44, 155–156
  tic-tac-toe, 34
Alphabet, 96
AlphaGo, 33–37
Amazon, 115, 158
Analytical engine, 76
Anarcho-capitalism, 83
Anderson, C. W., 46–47
Angwin, Julia, 154–156
App hackathons, 165–174
Apple Watch, 157

Artificial intelligence (AI)
  beginnings, 69–73
  expert systems, 52–53, 179
  fantasy of, 132
  in film, 31, 32, 198
  foundations of, 9
  future of, 194–196
  games and, 33–37
  general, 10–11, 32
  narrow, 10–11, 32–33, 97
  popularity of, 90
  real vs. imagined, 31–32
  research, women in, 158
  sentience challenge in, 129
Asimov, Isaac, 71
Assembly language, 24
Association for Computing Machinery
    (ACM), 145
Astrolabe, 76
Asymmetry, positive, 28
Automation technology, 176–177
Autopilot, 121
Availability heuristic, 96

Babbage, Charles, 76–77
Bailiwick (Broussard), 182–185, 190–
    191, 193
Barlow, John Perry, 82–83
Bell Labs, 13
Bench, Shane, 84

Ben Franklin Racing Team (Little Ben), 122–127
Berkman Klein Center (Harvard), 195
Berners-Lee, Tim, 4–5, 47
Bezos, Jeff, 73, 115
Bias
  in algorithms, 44, 150, 155–157, 195
  in algorithms, racial, 44, 155–156
  genius myth and, 83–84
  programmers and, 155–158
  in risk ratings, 44, 155–156
  in STEM fields, 83–84
Bill & Melinda Gates Foundation, 60–61, 157
Bipartisan Campaign Reform Act, 180
Bitcoin, 159
Bizannes, Elias, 165, 166, 171
Blow, Charles, 95
Boggs, David, 67–68
Boole, George, 77
Boolean algebra, 77
Borden, Brisha, 154–155
Borsook, Paulina, 82
Bowhead Systems Management, 137
boyd, danah, 195
Bradley, Earl, 43
Brains 19–20, 95, 128–129, 132, 144, 150
Brand, Stewart, 5, 29, 70, 73, 81–82
Brin, Sergei, 72, 151
Brown, Joshua D., 140, 142
Bump, Philip, 186
Burroughs, William S., 77
Burroughs, William Seward, 77

Calculation vs. consciousness, 37
Cali-Fame, 186
California, drug use in, 158–159
Cameron, James, 95
Campaign finance, 177–186, 191
Čapek, Karel, 129
Caprio, Mike, 170–171

Carnegie Mellon University, autonomous vehicle research
ALVINN, 131
  University Racing Team (Boss), 124, 126–127, 130–131
Cars
  deaths associated with, 136–138, 146
  distracted driving of, 146
  human-centered design for, 147
Cars, self-driving
  2005 Grand Challenge, 123–124
  2007 Grand Challenge, 122–127
  algorithms in, 139
  artificial intelligence in, 129–131, 133
  deaths in, 140
  driver-assistance technology from, 135, 146
  economics of, 147
  experiences in, 121–123, 125–126, 128
  fantasy of, 138, 142, 146
  GPS hacking, 139
  LIDAR guidance system, 139
  machine ethics, 144–145, 147
  nausea in, 121–123
  NHTSA categories for, 134
  problems/limitations, 138–140, 142–146
  research funding, 133
  SAE standards for levels of automation, 134–135
  safety, 136–137, 140–142, 143, 146
  sentience in, 132
  Uber's use of, 139
  Udacity open-source car competition, 135
  Waymo technology, 136
CERN, 4–5
Cerulo, Karen A., 28
Chess, 33
Children's Online Privacy Protection Act (COPPA), 63–64
Chinese Room argument, 38
Choxi, Heteen, 122

Christensen, Clayton, 163
Chrome, 25, 26
Citizens United, 177, 178, 180
Clarke, Arthur C., 71–72
Client-server model, 27
Clinkenbeard, John, 172
Cloud computing, 26, 52, 196
Cohen, Brian, 56–57
Collins, John, 117
Common Core State Standards, 60–61
Communes, 5, 10
Computer ethics, 144–145
Computer Go, 34–36
Computers
  assumptions about vs. reality of, 8
  components, identifying, 21–22
  consciousness, 17
  early, 196–199
  human, 77–78, 198
  human brains vs., 19–20, 128–129,
    132, 144, 150
  human communication vs., 169–170
  human mind vs., 38
  imagination, 128
  limitations, 6–7, 27–28, 37–39
  memory, 131
  modern-day, development of, 75–79
  operating systems, 24–25
  in schools, 63–65
  sentience, 17, 129
Computer science
  bias in, 79
  ethical training, 145
  explaining the world through, 118
  women in, 5
Consciousness vs. calculation, 37
Constants in programming, 88
Content-management system (CMS), 26
Cooper, Donna, 58
Copeland, Jack, 74–75
Correctional Offender Management
    Profiling for Alternative Sanctions
    (COMPAS), 44, 155–156

Cortana, 72
Counterculture, 5, 81–82
Cox, Amanda, 41–42
Crawford, Kate, 194
Crime reporting, 154–155
CTB/McGraw-Hill, 53
Cumberbatch, Benedict, 74
Cyberspace activism, 82–83

DarkMarket, 159
Dark web, 82
Data
  on campaign finance, 178–179
  computer-generated, 18–19
  defined, 18
  dirty, 104
  generating, 18
  people and, 57
  social construction of, 18
  unreasonable effectiveness of, 118–
    119, 121, 129
Data & Society, 195
DataCamp, 96
Data density theory, 169
Data journalism, 6, 43–47, 196
Data Journalism Awards, 196
Data journalism stories
  cost-benefit of, 47
  on inflation, 41–42
  Parliament members' expenses, 46
  on police speeding, 43
  on police stops of people of color, 43
  price discrimination, 46
  on sexual abuse by doctors, 42–43
Data Privacy Lab (Harvard), 195
Data Recognition Corporation (DRC), 53
Datasets in machine learning, 94–95
Data visualizations, 41–42
Deaths
  distracted driving accidents, 146
  from poisoning, 137
  from road accidents, 136–138
  in self-driving cars, 140

Decision making
  computational, 12, 43, 150
  data-driven, 119
  machine learning and, 115–116,
    118–119
  subjective, 150
Deep Blue (IBM), 33
Deep learning, 33
Defense Advanced Research Projects
    Agency (DARPA) Grand Challenge,
    123, 131, 133, 164
Desmond, Matthew, 115
Detroit race riots story, 44
Dhondt, Rebecca, 58
Diakopoulos, Nicholas, 46
Difference engine, 76
Differential pricing and race, 116
Digital age, 193
Digital revolution, 193–194
Dinakar, Karthik, 195
Django, 45, 89
DocumentCloud, 52, 196
Domino's, 170
Drone technology, 67–68
Drug marketplace, online, 159–160
Drug use, 80–81, 158–160
Duncan, Arne, 51
Dunier, Mitchell, 115

Edison, Thomas, 77
Education
  change, implementing in, 62–63
  Common Core State Standards, 60–61
  competence bar in, 150
  computers in schools, 63–65
  equality in, 77–78
  funding, 60
  supplies, availability of, 58
  technochauvinist solutions for, 63
  textbook availability, 53–60
  unpredictability in, 62
18F, 178–179
Electronic Frontier Foundation, 82

Elevators, 156–157
Eliza, 27–28
Emancipation Proclamation, 78
Engelbart, Doug, 25, 80–81
Engineers, ethical training, 145
ENIAC, 71, 194, 196–199
Equality
  in education, 77–78
  techno hostility toward, 83
  technological, creating, 87
  technology vs., 115, 156
  for women, 5, 77–78, 83–85, 158
Essa, Irfan, 46
Ethics, 144–145, 147
EveryBlock, 46
Expertise, cognitive fallacies associated,
    83
Expert systems, 52–53, 179

Facebook, 70, 83, 152, 158, 197
Facial recognition, 157
Fact checking, 45–46
Fake news, 154
Family Educational Rights and Privacy
    Act (FERPA), 63–64
FEC, McCutcheon v., 180
FEC, Speechnow.org v., 180
FEC.gov, 178–179
Film, AI in, 31, 32, 198
FiveThirtyEight.com, 47
Foote, Tully, 122–123, 125
Ford Motor Company, 140
Fowler, Susan, 74
Fraud
  campaign finance, 180
  Internet advertising, 153–154
Free press, role of, 44
Free speech, 82
Fuller, Buckminster, 74
Futurists, 89–90

Games, AI and, 33–37
Gates, Bill, 61

Gates, Melinda, 157–158
Gawker, 83
Gender equality, hostility toward, 83
Gender gap, 5, 84–85, 115, 158
Genius, cult of, 75
Genius myth, 83–84
Ghost-in-the-machine fallacy, 32, 39
Giffords, Gabby, 19–20
GitHub, 135
Go, 33–37
Good Old-Fashioned Artificial
    Intelligence (GOFAI), 10
Good vs. popular, 149–152, 160
Google, 72
Google Docs, 25
Google Maps API, 46
Google Street View, 131
Google X, 138, 151, 158
Government
    campaign finance, 177–186, 191
    cyberspace activism, antigovernment
        ideology, 82–83
    tech hostility toward, 82–83
Graphical user interface (GUI), 25, 72
Greyball, 74
Guardian, 45, 46

Hackathons, 165–174
Hackers, 69–70, 82, 153–154, 169, 173
Halevy, Alon, 119
Hamilton, James T., 47
Harley, Mike, 140
Harris, Melanie, 58–59
Harvard, Andrew, 184
Harvard University
    Berkman Klein Center, 195
    Data Privacy Lab, 195
    mathematics department, 84
"Hello, world" program, 13–18
Her, 31
Hern, Alex, 159
Hernandez, Daniel, Jr., 19
Heuristics, 95–96

Hillis, Danny, 73
Hippies, 5, 82
HitchBOT, 69
Hite, William, 58
Hoffman, Brian, 159
Holovaty, Adrian, 45–46
Home Depot, 46, 115, 155
Hooke, Robert, 88
Houghton Mifflin Harcourt (HMH),
    53
HP, 157
Hugo, Christoph von, 145
Human-centered design, 147, 177
Human computers, 77–78, 198
Human error, 136–137
Human-in-the-loop systems, 177, 179,
    187, 195
Hurst, Alicia, 164

Illinois quarter, 153–154
Imagination, 89–90, 128
Imitation Game, The (film), 74
Information industry, annual pay, 153
Injury mortality, 137
Innovation
    computational, 25
    disruptive, 163, 171
    funding, 172–173
    hackathons and, 166
Instacart, 171
Intelligence in machine learning,
    92–93
Interestingness threshold, 188
International Foundation for Advanced
    Study, 81
Internet
    advertising model, 151
    browsers, 25, 26
    careers, annual pay rates, 153
    core values, 150
    drug marketplace, 159–160
    early development of the, 5, 81
    fraud, 153–154

Internet (cont.)
  online communities,
    technolibertarianism in culture of,
    82–83
  rankings, 72, 150–152
Internet Explorer, 25
Internet pioneers, inspiration for, 5,
  81–82
Internet publishing industry, annual
  pay, 153
Internet search, 72, 150–152
Ito, Joi, 147, 195

Jacquard, Joseph Marie, 76
Java, 89
JavaScript, 89
Jobs, Steve, 25, 70, 72, 80, 81
Jones, Paul Tudor, 187–188
Journalism. *See also* Data journalism
  algorithmic accountability reporting,
    7, 43–44, 65–66
  artificial intelligence for, 52–53
  computational, 7, 46–47, 190
  computer-assisted reporting, 44–45
  machine learning in, 52
  precision reporting, 44
  social science applied to, 44

Kaggle, 96
Kalanick, Travis, 74, 139
Kamal, Fawzi, 139
Karel the Robot, 129–130
Karpathy, Andrej, 149
Kay, Alan, 25, 72
Ke Jie, 33
Kernighan, Brian, 13
Kesey, Ken, 81
Kilgore, Barney, 152
Kinect, 157
Kleinberg, Jon, 155–156
Krafcik, John, 136, 137
Kroeger, Brooke, 78
Kubrick, Stanley, 71

Kunerth, Jeff, 43
Kurzweil, Raymond, 73, 89, 90
Kushleyev, Alex, 124–125

Language
  computational communication
    problems, 87–89
  fluidity of, 91
  human vs. mathematical, 88
  naming problem in, 88–89
Lanier, Jaron, 145–146
Lazer, David, 115
Leadership gender gap, 158
Learning, 89
LeCun, Yann, 90
Lee, Dan, 123
Leibniz, Gottfried, 75, 76, 77
Lench, Heather, 84
Leslie, S. J., 83
Lessig, Lawrence, 194
Levandowski, Anthony, 139–140,
  142
Levy, Steven, 70
Lexus, 123, 140
Libertarianism, 82–83, 138, 159–160
Libraries, 96–97
Lightoller, Charles, 116
Lincoln, Abraham, 78
LinkedIn, 158
Linux, 24–25
Lipton, Zachary, 114
Literacy, technological, 21
Long, Milton, 117–118
Long Now Foundation, 73
Lord, Walter, 117–119
Loughner, Jared Lee, 19
Lovelace, Ada, 76
LSD, 81
Lucas, George, 70

Machine intelligence, determining,
  37–38
Machine language, 24

Machine learning
  algorithms in, 94
  defined, 11, 89, 91–92
  doing DataCamp *Titanic* tutorial,
    96–119
  datasets in, 94–96
  intelligence in, 92–93
  in journalism, 52
  limitations, 119
  linguistic confusion, 89
  reinforcement, 93
  social decision making and, 115–116,
    119
  supervised, 93
  training data in, 93–94
  unsupervised, 93
Machines, intelligence in, 33
Mahfouz, Christl, 186
Mapping, digital, 131
Masch, Michael, 57
Mathematical lookup tables, 77
Mathematics
  cult of genius, worship of, 75
  developing machines for, 75–79
  gender gap, 84–85
  gender stereotypes associated with, 84
  social culture of, 75
  women in, 77–78
May, Patrick, 158
McCarthy, John, 70, 71
McCutcheon v. FEC, 180
McIntyre, Tim, 170
McNamee, Roger, 138
Mercedes, 144
Merideth, Willie, 68
Meyer, Philip, 43
Microsoft, 25, 157
Minimum viable product (MVP),
    189–190
Minsky, Margaret, 79
Minsky, Marvin, 69–75, 79–80, 81, 84,
    89, 129, 132, 145, 193
MIT Artificial Intelligence Lab, 70

Mitchell, Tom M., 92
MIT Media Lab, 70, 72, 195
MIT Tech Model Railroad Club (TMRC),
    69–70
Models, mathematical, 94
Monty Python, 89
Moore School of Engineering, 196–198
Morais, Betsy, 167
Mortensen, Dennis, 132
Motor vehicle traffic-related injury
    mortality, 137–138
Mullainathan, Sendhil, 155–156
Munro, Randall, 87
Murdoch, William, 116
Musk, Elon, 142, 143–144

Naming problem, 88–89
Natanson, Hannah, 84
National Highway Traffic Safety
    Association (NHTSA), 133–134
Natural resources homework, 51–52
Navy, U.S., 137
Neumann, John von, 71
Neural networks, 33
Neville-Neil, George V., 92–93
New Communalism movement, 5
Newman, Barry, 152
Newspapers, 152
New York Times, 152
NeXt cube, 5
NICAR conference, 196
Nineteenth Amendment, 78
Northpointe, 155–156
Norvig, Peter, 93, 118
Nutter, Michael, 53
NVIDIA, 140–141

Obama Administration, 147, 194
Object, 97
O'Neil, Cathy, 94
One Laptop Per Child (OLPC) initiative,
    65
oN-Line System (NLS), 25

OpenBazaar, 159
Operating systems, 24–25
Opioid crisis, 158–160
O'Reilly, Tim, 81
OSX, 25
Otto, 142
Overview Project, 52

Page, Carl Victor, Sr., 72–73
Page, Larry, 72–73, 131, 151
PageRank, 72, 151–152
Palantir, 83
Panama Papers, 196
Pandas library, 97
Paperclip theory, 89–90
Papert, Seymour, 72, 73
Pasquale, Frank, 115
Pattis, Richard, 129
PayPal, 83, 159
Pearson, 53–54
Penn and Teller, 70
Pennsylvania System of School
    Assessment (PSSA), 52, 53–54
Pereira, Fernando, 118
Personal computer revolution, 5, 24
Philadelphia School District, 53–60,
    65–66
Physicians, sexual abuse by, 42–43
PillyPod, 173
Pinker, Stephen, 90
Pinkerton, Emma, 164
Pizzafy, 165, 168–174
Policing
    quantitative methods to enhance, 155
    racial disparities found by Stanford
        Open Policing Project, 43
    speeding, 43
PolitiFact, 45–46
Popular vs. good, 149–152, 160
Poverty and differential pricing, 116
Prater, Vernon, 155
Predictive analytics, 33
Price discrimination, 46

Price optimization, 114–115
Privacy, right to, 68, 195
Programmers
    accountability for, 154
    bias, 155–158
    competence, developing, 169–170, 174
    drug use, 158–159
    ethical training, 145
    income, 170–171
    safety, attitudes toward, 73–74
    social conventions, 74–75
Programming. See also Software
    anticipating the unexpected in, 28
    artisanal small-batch model, 191
    assembly language, 24
    command line, 15
    "Hello World," 13–18
    Karel the Robot exercise, 129–130
    modularity in, 17–18
    variables in, 88
    wealth, impact on, 158
    WHILE loops, 16
ProPublica, 45
Ptolemy, 77
Public transportation, 138
Pulitzer Prizes, 45
Punch-card loom, 76
Python, 14–17, 89, 92, 96

Race and image recognition systems,
    157
Radiation safety, 73–74
Radio broadcasting industry, annual
    pay, 153
Raghavan, Manish, 155–156
Rahwan, Iyad, 195
Reddit, 82
Reinhardt, Django, 89
Repetti, Steve, 166, 171
Reporting. See also Journalism
    computer-assisted, 44–45
    precision, 44
    robot, 9–10

Richardson, Kathleen, 72
Risk, drawing conclusions about, 95–96
Risk algorithms, 44–45, 155–156
Ritchie, Dennis, 13, 24
Road accident statistics, 136–138
Roberts, Jerry, 74–75
Robinett, Ricky, 171
Robot-car technology, 123. *See also* Cars, self-driving
Robot reporting, 9–10
Robots, 3–4, 87–88, 129
Rogers, Edwin, 164–165, 167, 172, 173
Roomba, 88
Royal, Cindy, 47
Rudisch, Gloria, 79
Russell, Stuart, 93

Safari, 25, 26
Science fiction, 71–72
Scikit-learn, 92, 96
Screet, 172
Sculley, John, 72
Searle, John, 38
Selfies, 149
SendGrid, 168
Seneca Falls Convention, 78
Sentience
  computer, 17, 129
  in self-driving cars, 132
Sexual abuse, 42–43, 45
Sexual harassment, culture of, 74
Shar.ed, 171–172
Sharkey, Pat, 115
Shaw, Jennifer, 164–165, 167, 170–171, 173
Sheivachman, Andrew, 187
Shell programming language, 15
Siegelmann, Hava, 133
Silicon Valley, 166
Silk Road, 159
Silver, Nate, 47
Singh, Santokh, 137
Singularity theory, 89–90

Siri, 28, 29, 72
Slavery, 78
Slovic, Paul, 83
Smart games format (SGF), 35
Smith, Dre, 167, 172
Smith, Edward John, 117
Snowden documents, 196
Social decision making, 115–116, 119
Social media, 158
Society, impact of algorithmic accountability reporting, 65–66
Society-in-the-loop machine learning, 195
Software. *See also* Programming
  autonomous systems, 187
  in the cloud, 25–26
  defined, 19, 22
  development process problems, 190
  human-in-the-loop systems, 177, 179, 187, 195
  lifespan, 193
  minimum viable product, 190
  naming, 182
  scope creep, 61
  technical debt, 193
Somerville, Heather, 158
Space elevator, 71–72
SPACES, 172–173
Speechnow.org v. FEC, 180
Spence, Stephen, 58–59
Standardized testing, 52–55
Stanford Racing Team (Junior), 124, 130–131
Stanford Racing Team (Stanley), 123–124, 127
Staples, 46, 115
Star Trek: The Next Generation, 31
Startup Bus, 163–174
Startup House, 166
Steiger, Paul, 45
STEM fields, 5, 83–85, 158
Step reckoner, 76
Stewart, Alex, 122, 125–126

Story Discovery Engine (Broussard), 178–180, 187, 188–191
Survivor (television), 164
Sweeney, Latanya, 195

Tacocopter, 29–30
Taplin, Jonathan, 83
Tay Twitter bot, 69
Teachers, underground economy, 57
TechCrunch Disrupt hackathon, 166–167
Tech culture
  drug use in, 158–160
  misogyny in, 167
  money in, 171
Tech Model Railroad Club (TMRC), 69–70
Technochauvinism
  assumptions from, 156
  beliefs accompanying, 8
  blaming drivers, 136
  defined, 7–8
  disruptive innovation and, 163
  hallmarks of, 69
  magical thinking of, 122
  philosophical basis of, 75
Technolibertarianism, 82–83
Technology
  breakage, 63, 156–157, 193
  digital, uses for, 194
  equality, creating, 87
  gender gap, 158
  human-centered design, 177
  inclusive, need for, 154
  inequality and, 83, 115, 156
  libertarianism and, 82–83
  limitations, 6–7, 176–177
  mathematical, development of, 75–79
  promises of, questioning the, 6
  social consequences, negative, 67–69
  white male bias in, 72, 79
Terminal, 14–15
Tesla, 121, 136, 139, 140–144
Textbooks, 53–60, 63–65

Texting while driving, 146–147
Thayer, Jack, 117–118
Thiel, Peter, 83, 159
Thirteenth Amendment, 78
Thrun, Sebastian, 124, 131, 135, 138
Tic-tac-toe, 33–34
Tilden Middle School, 59–60
Titanic (disaster), 95–119
Titanic (movie), 95
Torvalds, Linus, 24
Toyota, 140
Trolley problem, 144, 147
Trump, Donald, 83, 184–187, 194
Tufte, Edward, 169
Turing, Alan, 33, 74–75, 82, 83, 193
Turing test, 33, 37–38
Turner, Fred, 5, 81
Twitter, 69
2001: A Space Odyssey (film), 31, 71, 198

Uber, 74, 121, 138, 139–140, 142, 168
Udacity, 135, 138
Ukpeaġvik Iñupiat Corporation, 137
Ulbricht, Ross, 159
Unix, 24
Unmanned autonomous systems (UAS), 137
Urmson, Chris, 135
Usher, Nikki, 47

Vaporware, 166
Vehicles, SAE definitions for automated, 134–135. *See also* Cars, self-driving
Venmo, 159
Vernaza, Paul, 122, 126
Vicemo.com, 159
Video games, 25
Voice assistants, 28–29, 72
Voice interfaces, 38–39
Voting rights for women, 78

Waite, Matt, 45
Wall Street Journal, 45, 46, 152

Waydo, Jaime, 135–136
Waymo, 135–136, 139–140, 141
Web search portals industry, annual
    pay, 153
Web servers, 26–27
Whittaker, Meredith, 194
Whole Earth Catalog, 5, 73, 81–82
Whole Earth eLectronic Link (WELL),
    82
Wiesner, Jerry, 71
Wigner, Eugene, 118
Wilkinson, Jim, 75
Willis, Derek, 179
Wilson, Christo, 115
Winograd, Terry, 73
Wired, 29, 81
Wolfe, Tom, 81
Wolfram, Stephen, 79
Women's suffrage, 78
Worley, Cole, 171
Wright brothers, 3, 131–132
Writing, automated, 52

Xerox PARC, 25, 72

Yoshida, Junko, 135

Zaneski, Eddie, 168–169
Zimmerman, Henry, 71
Zuckerberg, Mark, 31, 70